Annually Laminated Lake Sediments

Annually Laminated Lake Sediments

Special Issue Editors

Wojciech Tylmann
Bernd Zolitschka

MDPI • Basel • Beijing • Wuhan • Barcelona • Belgrade • Manchester • Tokyo • Cluj • Tianjin

Special Issue Editors
Wojciech Tylmann
University of Gdańsk,
Institute of Geography
Poland

Bernd Zolitschka
University of Bremen,
Institute of Geography,
GEOPOLAR
Germany

Editorial Office
MDPI
St. Alban-Anlage 66
4052 Basel, Switzerland

This is a reprint of articles from the Special Issue published online in the open access journal *Quaternary* (ISSN 2571-550X) (available at: https://www.mdpi.com/journal/quaternary/special_issues/laminated_lake_sediments).

For citation purposes, cite each article independently as indicated on the article page online and as indicated below:

LastName, A.A.; LastName, B.B.; LastName, C.C. Article Title. *Journal Name* **Year**, *Article Number*, Page Range.

ISBN 978-3-03928-786-4 (Pbk)
ISBN 978-3-03928-787-1 (PDF)

Cover image courtesy of Wojciech Tylmann. It is a close-up of the thin section showing complex structure of varves from Lake Jaczno in northeastern Poland.

Contents

About the Special Issue Editors

Wojciech Tylmann (Prof. Dr.): A physical geographer interested in geomorphology, paleolimnology, and especially varved lake sediments with their fascinating potential for reconstructions of environmental changes. After completing his PhD thesis at the University of Gdańsk (Poland), he started exploring varved sediment records from northern Poland using sedimentological, geochronological, and geochemical methods. Recently, he has been involved in several projects that investigate the mechanisms of varve formation in lakes and the preservation of climate signals in lake sediments. Since the year 2014, he has been a Professor of Geography at the University of Gdańsk where he leads the Environmental Change Reconstruction Laboratory.

Bernd Zolitschka (Prof. Dr.): After studying maar lake sediments for his diploma and PhD theses at the University of Trier (Germany), Bernd Zolitschka focused his research on high-resolution varved (annually laminated) sediment records as a tool for dating, high-resolution environmental reconstruction, and Quaternary paleoclimatology. As a postdoctoral scientist at the Climate Systems Research Center (University of Massachusetts, USA) and the GFZ German Research Centre for Geosciences in Potsdam, he investigated high-resolution lacustrine systems from the Mediterranean to the High Arctic. Since the year 2000, he has been Professor in Physical Geography at the University of Bremen (Germany), engaged in interdisciplinary and international research initiatives such as the International Continental scientific Drilling Project (ICDP) PASADO or the Northern Polish Lake Research (NORPOLAR).

quaternary

MDPI

Editorial

Annually Laminated Lake Sediments—Recent Progress

Wojciech Tylmann [1,*] and Bernd Zolitschka [2]

1 Faculty of Oceanography and Geography, University of Gdańsk, 80309 Gdańsk, Poland
2 Institute of Geography, University of Bremen, 28359 Bremen, Germany; zoli@uni-bremen.de
* Correspondence: wojciech.tylmann@ug.edu.pl

Received: 4 February 2020; Accepted: 27 February 2020; Published: 3 March 2020

check for updates

Abstract: The collection of papers entitled "Annually Laminated Lake Sediments" illustrates the recent progress made in varved sediment research and highlights the variety of methodological approaches and research directions used. The contributions cover the monitoring of modern sediment fluxes using sediment traps, geochronological and sedimentological analyses of varves, multi-proxy investigations, including geochemical and biological proxies, as well as spatiotemporal analyses based on multi-core studies supported by satellite images. The scientific issues discussed the influences of hydroclimatological phenomena on short-term changes in sediment flux, the relationships between biogeochemical processes in the water column and the formation of varves, the preservation of environmental signals in varves, and possibilities of synchronizing varved records with other high-resolution environmental archives.

Keywords: environmental monitoring; geochronology; hypoxia; pollen analysis; sediment flux; varve microfacies

---

Annually laminated (varved) sediments are excellent archives of the changing Earth. They store information with a high temporal resolution about past climate variability, natural rates of environmental change, and anthropogenic disturbances [1]. The fundamental advantage of varves is the possibility of establishing precise time scales along sediment cores, which provide a solid geochronological framework and builds an essential element of paleoenvironmental reconstruction. In addition to the calendar-year time control, another advantage of varves is the absence of bioturbation in these natural archives enabling the application of innovative sub-millimeter-scaled scanning techniques for the generation of high-resolution (annual, sub-annual or seasonal) proxy data series (e.g., [2–4]). However, the lacustrine environment is prone to influences from a variety of natural and human-related forcing factors, all of which influence lake systems directly (e.g., water temperature, mixing regime, water column oxygenation, biological productivity) and indirectly (e.g., vegetation cover in the catchment area, soil erosion with an influx of minerogenic particles, pollutants and nutrients) with different intensities and at different time intervals. Complex interactions within the lake-catchment system and their effects on the lacustrine deposition cause uncertainties in the interpretation of reconstructions based on proxy records from lake sediments. Disentangling these overlapping influences is one of the most difficult problems to tackle [5]. However, new findings of lakes with varved sediments [6,7] and rapid developments in monitoring and analytical techniques shed new light on the potential of varves for paleolimnological reconstructions.

The initiative for this Special Issue emerged from activities of the past global changes (PAGES) Varve Working Group that was established to coordinate activities leading to methodological developments and, at the same time, setting quality standards for varve chronologies, data management and processing, as well as integration with other high-resolution, i.e., annually resolved environmental archives [8]. The authors present recent progress reports for different directions in varved sediment research. This is illustrated by the diversity of papers involving different methods and applied techniques, as well as different lake settings and geographical locations.

In their work, Johansson et al. [9] showed a novel technique for monitoring modern sedimentation processes in Lake Kallavesi, eastern Finland. They track the direct influence of meteorological and hydrological phenomena on short-term changes in sediment fluxes using the prototype of a high-resolution online sediment trap to obtain flux-rate information with daily resolution. In addition to typical seasonal variability, this research also provides valuable information on catchment responses to short-term and weather-related events, e.g., influences of storm tracks that cause larger sediment fluxes over several days. This new technique of sediment trapping combined with microstratigraphic analyses of varved sediment archives helps in the interpretation of paleoclimatic proxy records and improves the modeling of past weather and erosion conditions.

Salminen et al. [10] documented a multi-core investigation of spatiotemporal variations in hypolimnetic hypoxia for the past 200 years in Lake Lehmilampi, eastern Finland. They used spatiotemporal changes in varve distribution as an indicator for hypolimnetic hypoxia oscillations. This approach allowed for the recognition of several periods with hypoxia primarily related to changing climatic conditions. They conclude that more sophisticated investigations are essential for a better understanding of climatic and anthropogenic forcing behind hypolimnetic hypoxia oscillations in boreal lakes.

The potential of high-resolution pollen analysis in varve records from northern Germany was presented by Theuerkauf et al. [11]. They test whether pollen analysis with annual resolution can be used for synchronizing different varve records. The results for species that show strong annual variations in flowering, e.g., *Fagus sylvatica* and *Picea abies*, supported their chronostratigraphic potential and indeed allowed to synchronize these records. Moreover, peaks in pollen deposition correlate with minima in tree-ring widths, which provides a tool for the synchronization of these records.

Thys et al. [12] investigated the potential of sediments from Kenai Lake in south-central Alaska to archive snow avalanches. Using a precise chronology, sedimentological characteristics of varves, and satellite images, they explain how sediments are redistributed by ice floes across the lake and compare dropstone records with climate data. They present a direct link between historical snow avalanches and dropstones preserved in lake sediments, which suggests that dropstones can be used for long-term reconstructions of snow-avalanches with at least decadal resolution.

Additionally, Vegas-Vilarrúbia et al. [13] focused on the preservation of different environmental signals in varves. They synthesize results of different modern analogue studies performed for Lake Montcortès, Central Pyrenees (Spain), and discuss different limnological and biogeochemical processes that reveal seasonality in the lake and indicate different potentials for preservation in the sublayers of each varve. Advantages and limitations of such a multiproxy modern-analogue approach for investigations of lake sediments are discussed, and guidelines for future research at Lake Montcortès and beyond are proposed.

Finally, classical microstructural investigations of varves and their components were presented by Żarczyński et al. [14], who compare varve microfacies and particle-size distributions in biogenic varves of Lake Żabińskie in northeastern Poland. Although changes in particle-size distributions between different varve types were relatively small, end-member analyses allowed recognizing variabilities in the deposition of allochthonous and autochthonous sediment components. This suggests that there is room for more detailed sedimentological investigations not only in clastic but also in biogenic depositional environments. Grain size analysis of individual sediment components is suggested as a future task to improve the interpretation of end-member analysis.

All the papers in this volume demonstrate that annually laminated (varved) lake sediments are reliable recorders of past environmental change. However, the studies also highlight the complexity of the information embedded in these natural archives. Further investigations are required to improve our understanding of varves and their fascinating potential for reconstructions of natural and man-made changes in the environment, especially in modern times of ongoing global climate change.

Funding: This research received no external funding.

Acknowledgments: We sincerely thank Arndt Schimmelmann for improving an earlier version of this manuscript and the editors of the journal "*Quaternary*" for making this special issue possible.

Conflicts of Interest: The authors declare no conflict of interest.

References

1. Zolitschka, B.; Francus, P.; Ojala, A.E.K.; Schimmelmann, A. Varves in lake sediments—A review. *Quat. Sci. Rev.* **2015**, *117*, 1–41. [CrossRef]
2. Butz, C.; Grosjean, M.; Fischer, D.; Wunderle, S.; Tylmann, W.; Rein, B. Hyperspectral imaging spectroscopy: A promising method for the biogeochemical analysis of lake sediments. *J. Appl. Remote Sens.* **2015**, *9*, 096031. [CrossRef]
3. Rothwell, R.G.; Croudace, I. *Micro-XRF Studies of Sediment Cores: Perspective on Capability and Application in the Environmental Sciences*, 1st ed.; Springer: Dordrecht, The Netherlands, 2015.
4. Croudace, I.W.; Löwemark, L.; Tjallingii, R.; Zolitschka, B. Current perspectives on the capabilities of high resolution XRF core scanners. *Quat. Int.* **2019**, *514*, 5–15. [CrossRef]
5. Mills, K.; Schillereff, D.; Saulnier-Talbot, E.; Gell, P.; Anderson, N.J.; Arnaud, F.; Dong, X.; Jones, M.; McGowan, S.; Massaferro, J.; et al. Deciphering long-term records of natural variability and human impact as recorded in lake sediments: A palaeolimnological puzzle. *WIREs Water* **2017**, *4*, e1195. [CrossRef]
6. Ojala, A.E.K.; Francus, P.; Zolitschka, B.; Besonen, M.; Lamoureux, S.F. Characteristics of sedimentary varve chronologies—A review. *Quat. Sci. Rev.* **2012**, *43*, 45–60. [CrossRef]
7. Tylmann, W.; Zolitschka, B.; Enters, D.; Ohlendorf, C. Laminated lake sediments in northeast Poland: Distribution, preconditions for formation and potential for paleoenvironmental investigation. *J. Paleolimnol.* **2013**, *50*, 487–503. [CrossRef]
8. Zolitschka, B.; Pike, J.; von Gunten, L.; Kiefer, T. Annual Recorders of the Past. *Past Glob. Chang. Mag.* **2014**, *22*, 1–56. [CrossRef]
9. Johansson, M.; Saarni, S.; Sorvari, J. Ultra-High-Resolution Monitoring of the Catchment Response to Changing Weather Conditions Using Online Sediment Trapping. *Quaternary* **2019**, *2*, 18. [CrossRef]
10. Salminen, S.; Haltia, E.; Saarni, S.; Fukumoto, Y.; Saarinen, T. Varve distribution reveals spatiotemporal hypolimnetic hypoxia oscillations during the past 200 years in Lake Lehmilampi, Eastern Finland. *Quaternary* **2019**, *2*, 20. [CrossRef]
11. Theuerkauf, M.; Engelbrecht, E.; Dräger, N.; Hupfer, M.; Mrotzek, A.; Prager, A.; Scharnweber, T. Using annual resolution pollen analysis to synchronize varve and tree ring records. *Quaternary* **2019**, *2*, 23. [CrossRef]
12. Thys, S.; Van Daele, M.; Praet, N.; Jensen, B.J.L.; Van Dyck, T.; Haeussler, T.; Vandekerkhove, E.; Cnudde, V.; De Batist, M. Dropstones in lacustrine sediments as a record of snow avalanches—A validation of the proxy by combining satellite imagery and varve chronology at Kenai Lake (south-central Alaska). *Quaternary* **2019**, *2*, 11. [CrossRef]
13. Vegas-Vilarrúbia, T.; Rull, V.; del Carmen Trapote, M.; Cao, M.; Rosell-Melé, A.; Buchaca, T.; Gomà, J.; López, P.; Sigró, J.; Safont, E.; et al. Modern Analogue Approach Applied to High-Resolution Varved Sediments—A Synthesis for Lake Montcortès (Central Pyrenees). *Quaternary* **2020**, *3*, 1. [CrossRef]
14. Żarczyński, M.; Szmańda, J.; Tylmann, W. Grain-size distribution and structural characteristics of varved sediments from Lake Żabińskie (northeastern Poland). *Quaternary* **2019**, *2*, 8. [CrossRef]

quaternary

Article

Ultra-High-Resolution Monitoring of the Catchment Response to Changing Weather Conditions Using Online Sediment Trapping

Markus Johansson [1], Saija Saarni [2] and Jouni Sorvari [1,*]

[1] Department of Environmental and Biological Sciences, University of Eastern Finland, P.O. Box 1627, FI-70211 Kuopio, Finland; markus.johansson@uef.fi

[2] Department of Geography and Geology, University of Turku, FI-20014 Turku, Finland; saitur@utu.fi

* Correspondence: jouni.sorvari@uef.fi; Tel.: +358-40-355-3190

Received: 31 March 2019; Accepted: 6 May 2019; Published: 12 May 2019

Abstract: The purpose of this study was to identify relationships between meteorological and hydrological observations and sediment flux rate changes, in order to better understand catchment dynamics. The meteorological and hydrological observations included local air temperature, wind speed, water temperature, and ice cover, while the sediment flux rate was observed in the lake basin using a modified sediment trap technique. This study demonstrates the advantages of a new online methodology applied in conventional sediment trapping to obtain flux rate information with daily resolution. A prototype of a high-resolution online sediment trap was tested in Savilahti Bay, Lake Kallavesi, eastern Finland, during the period from 22 October 2017 to 6 October 2018. The daily resolutions of meteorological, hydrological, and sediment flux rate data were analyzed using statistical methods. The results indicate relationships between temperature, precipitation, wind speed, and sediment flux rate, but the urban site also showed erosional changes due to anthropogenic land use. Sediment flux ceased during winter season and spring floods were recorded as pronounced peaks in sediment flux, while the growing season showed generally higher sediment accumulation rates. This research also provides valuable information on the catchment response to short-term weather events. The influence of a storm led to larger sediment flux for several days. The importance of wind speed and frost formation on sedimentation, which has been difficult to address due to trap deployment times of typically several months, is now supported. Used together with varved sediment archives, online sediment trapping will facilitate the interpretation of paleoclimatic proxy records and modeling of detailed weather and erosion conditions that are related to climate change.

Keywords: sedimentation; flux rate; catchment dynamics; ultra-high resolution; environmental monitoring; novel technology methodology

1. Introduction

Annually laminated (varved) sediments have shown their importance in paleoclimatological and environmental studies [1–3]. A high temporal resolution up to subannual scale allows robust dating [1] as well as investigation of past hydroclimatic and temperature variability, even in a season-specific manner [4–7]. Varves, formed as a consequence of seasonal cycles in sediment availability and production [2], are highly dependent on catchment characteristics and environmental conditions [2,8]. This leads to their sensitivity in reflecting changes in climatic conditions over time in different climatic zones [9–13] and underlines the need to understand each varve record with respect to their individual catchment and climate characteristics [2,4]. Understanding the detailed mechanisms that mediate hydroclimatic changes in sedimentary records is critical for paleoclimatic and environmental interpretations [2,4,14,15]. Understanding of the process is further gained by using sediment trap monitoring, which reveals seasonal changes in sedimentary material in detail [14–18].

Sediment trapping is a classical technique used to acquire time series of sediment flux rates and variations in sedimentary composition in aquatic environments [14,15]. In order to investigate the sediment flux rate with respect to meteorological and hydrological effects, high-resolution observations from the study site are required. Using sequential sediment trap monitoring with seasonal deployment intervals, it has been possible to gain a detailed understanding of the seasonal controls on catchment processes, erosion, production, and sediment deposition [14–18]. However, investigating the importance of a single short-term meteorological event in sediment deposition using the sediment trap technique is difficult. Traditionally, sediment trap time series are collected with a resolution of a few months. Nowadays, higher-resolution trap monitoring is desired and datasets of monthly and biweekly resolution are typical [14,19,20]. Thus, the influences of short-term meteorological and hydrological events on flux rate changes are not recorded, but are addressed on the scale of weeks. Furthermore, wind and frost formation, which are obscured by the more dominant forcing factors, have been suggested to influence sediment accumulation [4,21–23], but so far have been addressed using indirect evidence and statistical approaches. Using seasonal trap monitoring, these questions are difficult to answer, because other erosional events, such as spring floods or numerous rainy days, have dominated the sedimentation. So far, according to our knowledge, no systematic attempts have been carried out to measure the changes in sediment accumulation during, for example, a single storm event. Therefore, a monitoring technique that can obtain ultra-high-resolution time series from flux rates in aquatic sites is desired.

Trapping methods can be divided into two categories: passive sampling traps and traps with sample bottles attached in an automatic revolving wheel. Passive trapping is a commonly used method in lake studies, with deployment intervals varying from a few weeks to several months. A revolving wheel can be used to sample with higher temporal resolution. However, due to the restricted number of collector tubes in a revolver, sampling longer time series with daily resolution is rare due to increased operating costs and effort required. The modern online technology applied to sediment trap monitoring would enable longer and higher-resolution time series of flux rate variability and therefore a more detailed understanding of the processes related to sediment accumulation with respect to anthropogenic land use or single meteorological events.

Computer tomography methods are noninvasive and nonintrusive ways to measure and image the interiors of objects. Typically, tomographic imaging applications are utilized in medical and industrial fields but can be applied in other fields where there is a need to obtain images inside objects. Computer tomographic imaging combines multiple projections of an object into a cross-sectional image. Projections are measured by scanning the object from different angles. Tomographic results are in two or three dimensions, depending on the measurement setup. Three-dimensional images describe the height, length, and volume of the object.

We applied tomographic imaging to sediment traps in order to measure volume changes inside the sediment trap collector tube and determine flux rate variations semicontinuously with daily time resolution. This paper presents a methodology for ultra-high-resolution online sediment trapping to investigate daily changes in sediment flux rate with respect to meteorological and hydrological observations. This technique can be applied to understand the importance of different meteorological and hydrological events in catchment erosion and sedimentation in detail.

2. Materials and Methods

2.1. Site Description

Lake Kallavesi is a large water body located in eastern Finland surrounding the city of Kuopio. It is located in the boreal vegetation zone, with annual mean temperature of +3 °C, and mean temperature of the coldest and warmest months is −10 °C (January) and +17 °C (July), respectively [24]. Annual precipitation is about 644 mm/year, of which approximately half is in the form of snow. The lake is covered by ice for about 6 months, typically from November to April [25].

The complex Lake Kallavesi basin was formed after the Weichselian ice sheet retreated more than 9500 years ago [26,27]. It includes several semiclosed basins detached from the main water body by glacial deposits such as eskers, but also by bedrock shear zones and Precambrian domes. One semiclosed basin, Savilahti Bay, was selected as the study site because it has several advantages. With the University of Eastern Finland campus located on its shore, Savilahti Bay was an ideal place to test the prototype of the sediment trap system because of the ease of access to the trap location in case the online data showed interruption, disconnection, or other problems. In addition to the ease of trap maintenance, the advantages are undisturbed sedimentation in the deepest part of the basin and, first and foremost, the meteorological station located on the shore to enable the evaluation of the importance of weather events in sediment accumulation. The surface area of the Savilahti Basin is 0.345 km^2 and the maximum water depth is 7 m (Figure 1). Savilahti Bay is detached from the main water body of Lake Kallavesi by two sills, reducing the water depth to 1 m in eastern and northern parts of the basin. Thus, the sedimentation of Savilahti Bay can be considered to mostly represent its catchment instead of the entirety of Lake Kallavesi. There are two narrow streams discharging to Savilahti Bay from the northeastern island and southwestern coast. The bay catchment is composed of coarse-grained material including quaternary sand deposits and sand moraines, former lacustrine fine-grained sediments, and bedrock outcrops. Savilahti Bay is located at 81 m a.s.l. and the northeastern side of the bay is sheltered by a high hill slope (140 m a.s.l.). At present there are human residences, road systems, and other infrastructure on the catchment in addition to boreal pine forests.

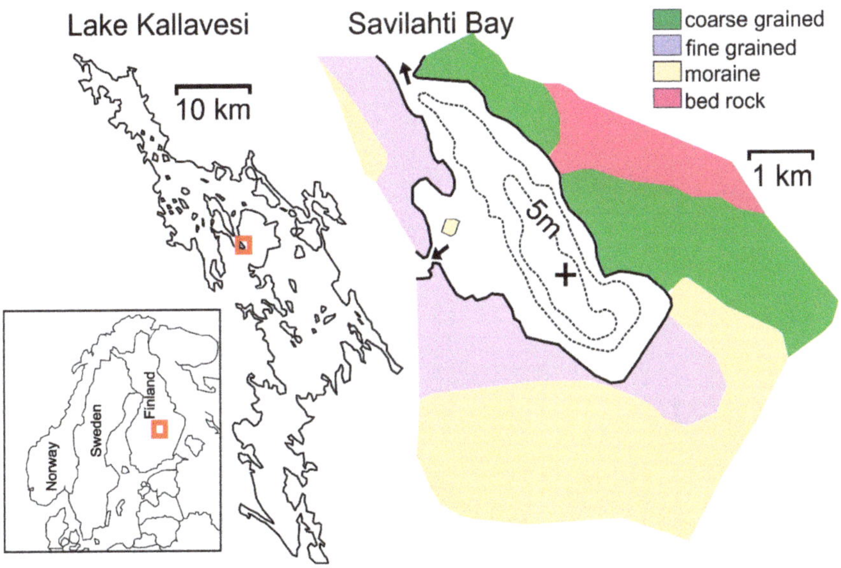

Figure 1. Location of Savilahti Bay and the online sediment trap site in the deepest part of the basin (marked with +).

Lake Kallavesi is a dimictic brown-water lake, which is a very typical lake type in Finland. Savilahti Bay suffers hypoxic conditions during the winter stratification. According to the measurements at the study site in February of 2016–2018, the oxygen concentration in bottom waters was 8.1, 4.2, and 4.4 mg/L, respectively. Thus, with its size, catchment, and water characteristics, Savilahti Bay resembles the previously reported Finnish lakes with annually laminated sediments [4,9,15,21,23].

2.2. Field Work

A prototype of a high-resolution online trap (Figures 2 and 3) was installed at the deepest point of the basin 1 m above the sediment surface, i.e., to a depth of 6 m. During the study, the sediment accumulation and the state of the equipment were inspected on daily basis. An anchored buoy that sends measurements over a mobile network was installed near the online trap (Figure 2a) and left in place for the duration of the study, including the period of ice cover. The daily resolution of meteorological and hydrological data was obtained using open data from the Finnish Meteorological Institute [24] and Finland's environmental administration [25]. Ice-cover information was based on visual observation. Sediment flux rate and meteorological and hydrological observations were collected during the period from 22 October 2017 to 6 October 2018 in Savilahti Bay of Lake Kallavesi, Kuopio, Finland. The sediment trap was collected on 11 November 2017, 10 May 2018, 28 July 2018, and 6 October 2018.

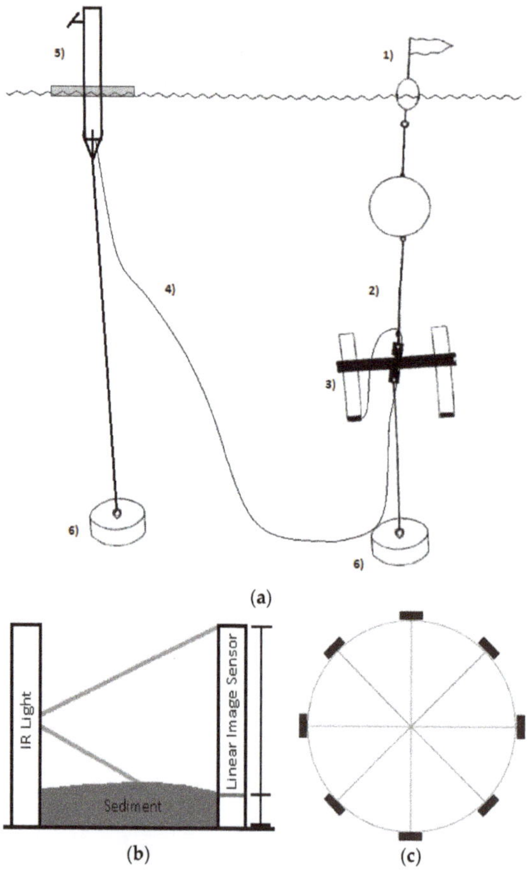

Figure 2. (**a**) Schematic illustration showing the online trap configuration: (1) signal buoy, (2) sediment trap, (3) trap collector tube with tomography device, (4) cable connecting tomography device with solar panel energy supply and Internet network, (5) float providing Internet connection and energy supply, and (6) anchors. (**b**) Cross-section illustration of tomographic scanning at the lower part of the modified trap collector tube, and (**c**) location of optical transmitters and receiver pairs at the bottom of the modified trap tube.

Figure 3. Online trap in Savilahti Bay basin at a depth of 6 m under the flag sign. Anchored float behind flag sign is the base station for the online trap, which sends measurements to the Internet over a mobile network.

2.3. Daily Measurement of Sediment Flux

Online trapping was used to investigate daily changes of sedimentation flux rate. A conventional sediment trap with 2 collector tubes was modified for this purpose. Computed tomography scanning equipment was adjusted in collector tube 1, while collector tube 2 was left as a reference tube (Figure 2a). Computed tomography, where the resulting image is a volumetric representation of structures in the trap tube, was applied to measure the daily change in sediment volume (Figure 2b). The trap tube was surrounded by 8 optical transmitter (infrared light) and receiver (linear camera) pairs in the inner surface at the lower part of the tube (Figure 2c). Optical transmitters and receivers performed tomographic scans of the trap tube interior. Computer tomography scans are based on optical measurements, which measure light attenuation from different angles. Accumulated sediment causes light attenuation by absorbing light energy. Water affects light transfer, but significantly less than accumulated sediment. This attenuation is captured by imaging sensors from opposite sides as light sources. Imaging sensors are positioned vertically to capture changes in height. Scanning was performed by turning light sources on and off sequentially and capturing light attenuation by imaging sensors simultaneously from the opposite side in a similar sequence. Captured light attenuation represents the light absorption profile of accumulated sediment from the height direction, i.e., projection of accumulated sediment. Profiles captured from 8 different angles were processed by a filtered back projection algorithm to a volumetric image. The measurement interval was 1 h, and data were transferred as an email attachment once a day. The operating principle and a detailed technical description of the online sediment trap methodology are provided in [28]. The measurement range in the height direction was 0 to 8.125 mm, and with a trap diameter of 56.5 mm, the maximum volume was about 19.9 mL. Accumulation was followed online so that the maximum volume would not be exceeded.

The sediment trap was collected 4 times during the year in order to verify the volume of accumulated sediment and empty the trap tube to ensure that the maximum technical volume was not exceeded. Accumulated sediment volume was measured manually from the trap tube in a measuring cylinder, in order to compare tomography scan volume to actual sediment volume in the tube. Sediment collection was performed by taking the whole contents of the trap, sediment and water, to a 5 L sample vessel. Collected sediment and water were stored in the sample vessel for further processing. To determine sediment volume, first the amount of water was reduced from the sample vessel to about 80 mL. Then, the sediment and remaining water were poured into a 100 mL measuring cylinder and left to settle for about 1 day. After settling, sediment volume was determined manually.

2.4. Meteorological and Hydrological Observations

Meteorological observations of open data from the Finnish Meteorological Institute were utilized [24]. The meteorological station is located only 200 m from the sediment trap. Meteorological data include air temperature (°C), air humidity (%RH), dew point (°C), precipitation (mm), pressure (hPa), wind speed (m/s), wind direction (°), snow depth (cm), visibility (m), water precipitation (mm), and snow precipitation (mm). The hydrological observations were from local measurements from the study site and an open data service provided by Finland's environmental administration. Local hydrological measurements included water temperature from 1–6 m depth within 1 m intervals. Depth is marked in meters in the results as 1–6, where 1 is near the water surface and 6 is near the lake bottom.

Hydrological data from Finland's environmental administration consisted of the following parameters: water surface temperature (°C) from a depth of 1 m, water level (cm), water flow (m^3/s), and depth of ground frost (cm). Ice cover information is based on visual observations of Savilahti Bay, recorded in the diary according to the situation in the bay. The ice cover data are interpreted as 0 for liquid water and 1 for ice cover. Ground frost data accuracy is about 1–2 cm; this specified accuracy is used to discriminate the ground frost variable to define 2 different situations. The ground frost variable is used as a grouping variable: 1 if the frost is more than 1 cm in depth and 0 if less than 1 cm. Ground frost group data are interpreted as unfrozen soil (0) and frost at least 1 cm deep (1).

Wind direction is a cyclic variable, which is transformed to vector values by Equations (1) and (2). Equation (1) is used to transform the wind direction to the north–south vector, and Equation (2) is used to transform the wind direction to the west–east vector:

$$\text{Wind direction (north-south)} = \text{wind speed m/s} \times \sin(\text{wind direction degrees}) \tag{1}$$

$$\text{Wind direction (west-east)} = \text{wind speed m/s} \times \cos(\text{wind direction degrees}) \tag{2}$$

2.5. Statistical Analysis

All data, sediment accumulation, and meteorological and hydrological observations were combined into one dataset for statistical analysis. Statistical analysis was processed with SPSS software version 25.0 [29]. First, data were preprocessed by setting outlier values to variables and then calculating daily averages for continuous variables from one-hour resolution data. Discrete variables of water and snow precipitation were also preprocessed by setting outlier values to variables but calculating the daily sum instead of averaging. Pearson's correlation coefficient and scatter plots were used to measure linear dependencies of variables. The combined daily resolution dataset row count was 350. The outlier values were set for sedimentation flux rate to values outside of 0.01 to 1.00 mL/day. The meteorological and hydrological measurements were normally distributed, therefore no outliers were excluded. The local water temperature dataset row count was 278 for depths of 1 to 6 m. Local water temperature data failed to record temperatures from 27 July 2018 due hardware failure. The rest of the rows (279 to 350) of water temperature to one-meter depth were replaced by water surface temperature data from Finland's environmental administration. The statistical analysis was performed using grouping information to separate the effect of ice cover for cases where meteorological events may affect sedimentation flux rate. As another way to study the influence of weather conditions on sedimentation, ice cover observations were used to separate the dataset for 2 different cases. Air temperature, snow depth, and all hydrological observations were examined for the whole dataset.

3. Results

The monitored sediment flux rate showed great daily variation. The measured sediment accumulation varied from 0 to 1.73 mL/day, the mean accumulation was 0.11 mL/day, and the median was 0.10 mL/day during the one-year test period. The total accumulation for the year was 39 mL, of which 3.65 mL accumulated during winter (November–March), 8.07 mL during spring (April–May), 18.38 mL during summer (June–August), and 8.89 mL during autumn (September–October; note that autumn accumulation is the combined sum from 22–31 October 2017 and 1 September to 6 October 2018).

The sediment volume was measured manually in order to verify the tomographic scan volume measurements. Sediment collection dates and sediment volumes (in mL) are shown in Table 1. The table shows sediment volume measured by the online trap and manually; the last column of the table shows differences between online and manually measured volume. Total accumulated sediment was 39 mL as measured by the online device and 40.8 mL by manual measurement.

Table 1. Sediment collection dates and volumes.

Sediment Collection Date	Online Measurement (mL)	Manual Measurement (mL)	Difference (mL)
11.11.2017	3.0	2.5	0.5
10.5.2018	6.9	6.5	0.4
28.7.2018	14.7	14.0	0.7
6.10.2018	14.4	17.8	−3.4

3.1. Ultra-High-Resolution Sediment Flux Rate with Meteorological and Hydrological Observations

The sedimentation flux rate and meteorological and hydrological observations represent seasonal variations over the investigated time period of one year. Measured variables, units, and descriptive statistics are shown in Table 2. Time series figures represent variations and relationships between variables: sedimentation flux rate with meteorological and hydrological observations (Figures 4–6). Time series figures were interpreted as different seasonal situations.

Table 2. Descriptive statistics of variables used in study.

Variable	Unit	N	Minimum	Maximum	Mean	Median	Std. Deviation
Sedimentation Flux rate	mL/day	350	0.00	1.73	0.11	0.10	0.23
Air temperature	°C	350	−21.10	25.37	5.02	13.27	10.87
Wind speed	m/s	350	0.88	5.83	2.49	2.40	0.90
Wind gust	m/s	350	1.56	11.18	4.70	4.45	1.77
Wind direction	°	350	58.83	350.92	199.33	191.52	70.73
Snow depth	Cm	350	0.00	93.88	22.41	50.93	29.75
Water precipitation	mm	350	0.00	37.00	1.11	2.10	3.58
Snow precipitation	mm	350	0.00	18.50	0.62	2.00	1.99
Water temperature at depth of:	°C						
1 m		350	0.06	24.70	7.47	14.40	8.54
2 m		278	0.07	18.81	4.97	1.32	6.44
3 m		278	0.06	16.58	4.36	1.56	5.24
4 m		278	0.06	13.70	3.89	1.73	4.19
5 m		278	0.13	12.41	3.63	2.07	3.38
6 m		278	0.12	10.27	3.37	2.56	2.67
Ground frost	Cm	183	0.00	5.00	1.00	1.40	1.43
Ice cover	0 (no), 1 (yes)	350	0	1	-	-	-
Construction	0 (no), 1 (yes)	350	0	1	-	-	-
Ground frost (group variable)	0 (no), 1 (yes)	350	0	1	-	-	-

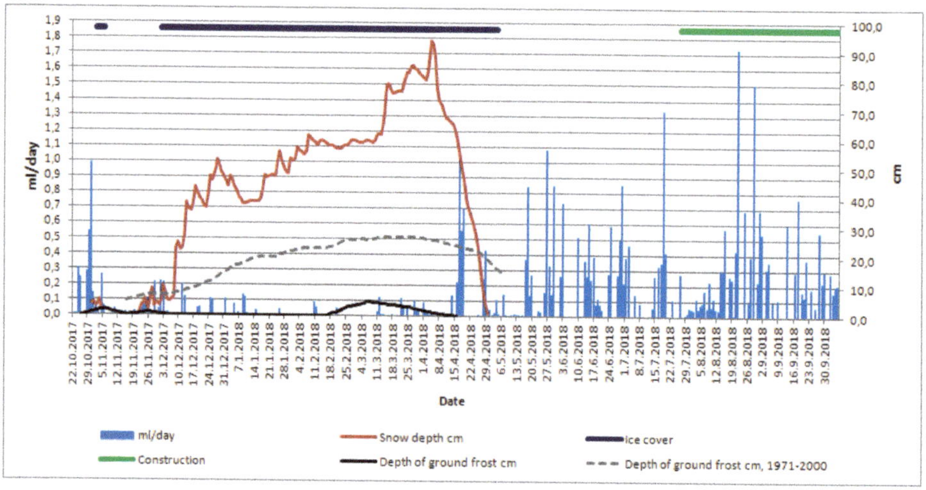

Figure 4. Daily sedimentation flux rate of Savilahti Bay from 22 October 2017 to 6 October 2018. Bar diagram shows sediment flux rate in mL/day, line diagram shows snow depth in cm (red line), ground frost depth in cm (black line), and monthly averaged ground frost depth from 1971 to 2000 in cm (dotted gray line). Note that ground frost in cm means depth below the ground surface. Bold lines on top of diagrams show ice cover (dark blue) and construction (green) events during the investigation period.

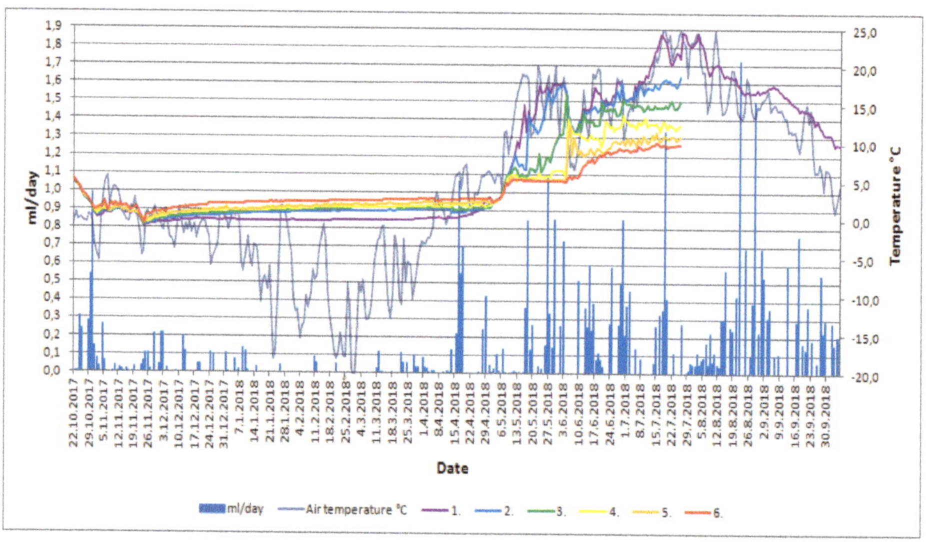

Figure 5. Daily sedimentation flux rate with respect to air temperature and basin water temperature at depths of 1 to 6 m, with one-meter interval, in Savilahti Bay from 22 October 2017 to 6 October 2018. Bar diagram shows sedimentation flux rate in mL/day and line diagram shows temperatures during investigation period: air (blue transparent line), water from 1 m depth (violet line), 2 m depth (blue line), 3 m depth (green line), 4 m depth (yellow line), 5 m depth (orange line), and 6 m depth (red line). Sedimentation flux rate increased when air temperature was above 0 °C and decreased when air temperature was below 0 °C. Note that water temperature from depths of 2 to 6 m are missing due hardware failure starting on 27 July 2018.

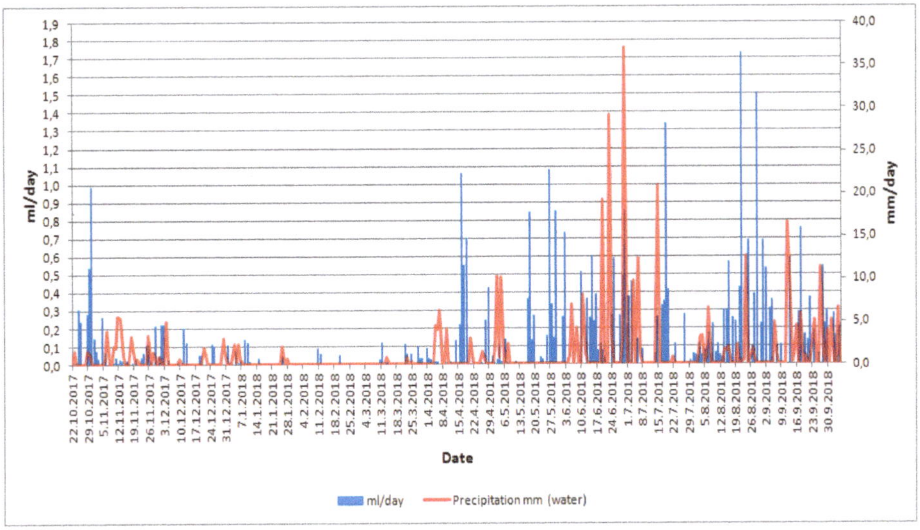

Figure 6. Daily sedimentation flux rate from Savilahti from 22 October 2017 to 6 October 2018. Bar diagram shows sedimentation flux rate in mL/day and line diagram shows water precipitation in mm/day (red line) during the investigation period.

3.2. Statistical Analysis Results

Pearson correlation coefficient was computed to assess the relationship between the sedimentation flux rate and meteorological and hydrological observations. Statistical correlations were calculated between every variable of the dataset, and statistically significant results where correlation coefficient *p*-value was <0.05 are shown in Tables 3 and 4.

Table 3. Pearson correlation between sediment flux rate and meteorological observations.

Flux Rate (mL/day)	Air Temperature (°C)	Wind Speed (m/s)	Wind Direction (West–East)	Precipitation (mm Water)
r	0.31 [1]	0.31 [2]	0.20 [2]	0.35 [1]
p	<0.001	0.02	0.045	0.01
N	350	54	106	54

[1] Correlation is significant at the 0.01 level (2-tailed). [2] Correlation is significant at the 0.05 level (2-tailed).

Table 4. Pearson correlation between sediment flux rate and hydrological observations.

Flux Rate (mL/day)	1	2	3	4
R	0.31 [1]	0.34 [1]	0.30 [1]	0.28 [1]
P	<0.001	0.02	0.01	<0.001
N	350	278	278	278
Flux Rate (mL/day)	**5**	**6**	**Ice Cover**	**Snow Depth (cm)**
R	0.28 [1]	0.28 [1]	−0.28 [1]	−0.26 [1]
P	0.01	<0.001	<0.001	<0.001
N	278	278	350	350

[1] Correlation is significant at the 0.01 level (2-tailed).

Comparing the Pearson correlation coefficient between sedimentation flux rate and meteorological observations (Table 3) shows a positive correlation between temperature, wind speed, wind direction, and water precipitation. The positive correlation between sediment flux rate and air temperature ($r = 0.31$, N = 350, $p < 0.001$) is summarized in a scatter plot (Figure 7a). Increases in sedimentation flux rate were correlated with increases in air temperature.

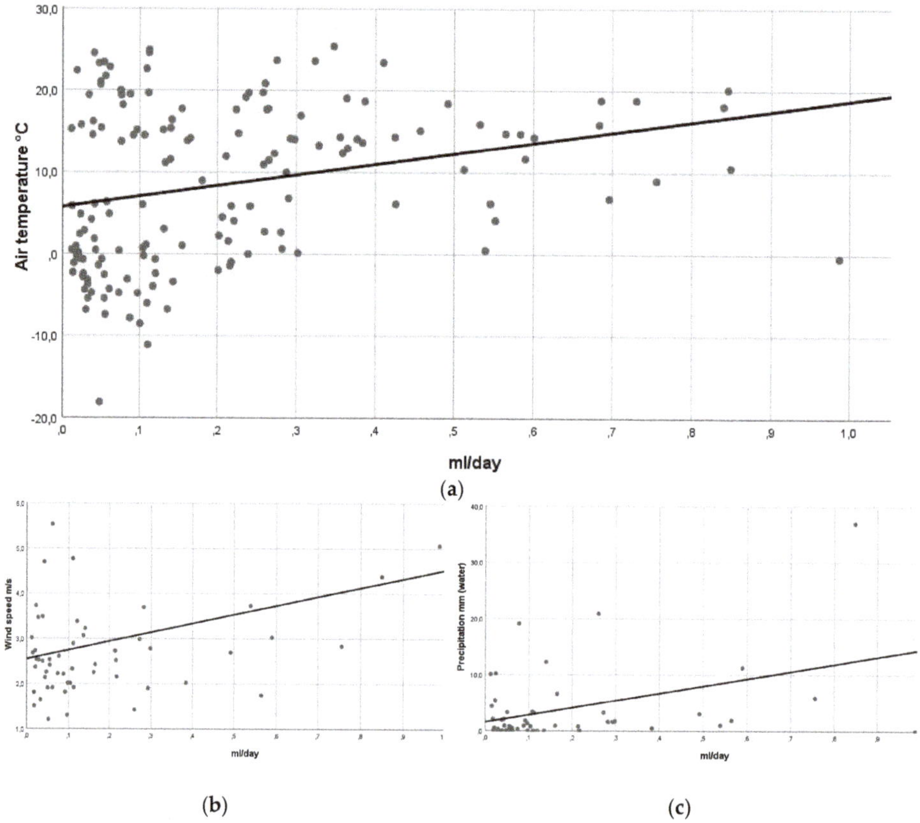

Figure 7. Linear relationship between (**a**) air temperature, (**b**) wind speed, and (**c**) water precipitation and daily sedimentation flux rate.

Overall, there was a medium positive correlation between sedimentation flux rate and wind speed ($r = 0.31$, N = 54, $p = 0.02$; Table 3, Figure 7b). Increased sediment flux was correlated with enhanced wind speed. A weak positive correlation between flux rate and wind direction was also found ($r = 0.20$, N = 106, $p = 0.045$), suggesting increased sedimentation during windy periods. Finally, there was a medium positive correlation between water precipitation and flux rate ($r = 0.35$, N = 54, $p = 0.01$). Increased sedimentation flux rate was correlated with increased water precipitation (Table 3, Figure 7c).

Positive and negative correlation coefficients between sedimentation flux rate and hydrological observations were found (Table 4). A medium negative correlation was observed between sediment flux rate and snow depth ($r = -0.26$, N = 350, $p < 0.001$). Increased sediment flux rate was correlated with decreased snow depth. Flux rate and water temperature at depths of 1–6 m showed a positive correlation (Table 4), while a medium negative correlation was found between sediment accumulation and occurrence of ice cover ($r = -0.28$, N = 350, $p < 0.001$). Sediment flux rate was decreased during periods of ice cover.

The 3-day precipitation event during 10–12 June with daily precipitation of 4.1 mm, 8.1 mm, and 1.2 mm, respectively, was followed by a significant increase in daily sediment accumulation on 13 June, which lasted up to 5 days. Single one-day storm events on 23 June, 30 June, and 15 July (see Figure 6) caused water precipitation of 29.2, 37.0, and 20.9 mm, respectively, followed by increased sediment accumulation on 24 June, 1 July, and 17 July with durations of 2, 4, and 4 days. It is noteworthy that in between precipitation events, the sediment flux rate decreased to a much lower level, even under the detection limit. The 6-day precipitation event during 10–15 November, with daily precipitation of 2.1 mm, 1.9 mm, 5.7 mm, 5.4 mm, 2.1 mm, and 0.3 mm, respectively, was followed by a minor increase in daily sediment accumulation.

The 2-day windy event without precipitation within 9 days on 18–19 May, with daily average wind speed (and wind gust) of 2.0 (3.9) and 3.9 (7.1) m/s in the SW and NW directions, respectively, was followed by a significant increase in daily sediment accumulation, which lasted up to 3 days. Another 5-day windy event without precipitation within 21 days on 27–31 May, with daily average wind speed (and wind gust) of 3.2 (6.4), 2.8 (5.2), 1.9 (3.8), 3.6 (7.7), and 4.5 (8.4) m/s in W, NW, W, W, and W directions, respectively, was followed by a significant increase in daily sediment accumulation, which lasted up to 6 days.

An independent-samples *t*-test was conducted to compare sedimentation flux rates in non-ice cover (group value 0) and ice cover (group value 1) conditions. There was a significant difference in the scores for non-ice cover (mean = 0.17, SD = 0.27) and ice cover (mean = 0.04, SD = 0.12) conditions ($t_{34} = 5.48$, $p < 0.001$). Our results show that when there was no ice cover, the sedimentation flux rate increased. An independent-samples *t*-test was conducted to compare sediment flux rates in non-ground frost (group value 0) and ground frost (group value 1) conditions. There was a significant difference in the scores for non-ground frost (mean = 0.23, SD = 0.21) and ground frost (mean = 0.07, SD = 0.06) conditions ($t_{15} = 3.01$, $p = 0.002$). These results suggest that the occurrence of ground frost has a negative effect on the sedimentation flux rate.

4. Discussion

4.1. Potential and Reliability of Online Sediment Trap

Multiple high-resolution paleoclimatological studies from varved sediments have largely increased our understanding of past climate changes and their drivers and scales. Several studies have shown the link between lacustrine sediment accumulation and large-scale climatic patterns such as solar activity [9,11,21,30,31] and changes in atmospheric circulation [9–12]. The detailed mechanisms by which catchments respond to large-scale climatic variability are very case-specific [2]. Detailed identification of the importance of specific weather and hydrological events will facilitate the interpretation of seasonal sedimentary records and improve proxy-based paleoclimatological records. Using online sediment traps with ultra-high (i.e., daily) resolution, it is possible to measure the variation of sediment fluxes in detail and develop our understanding of catchment processes and ultimately their climatic and hydrological controls.

In this study, we measured daily sediment accumulation flux rates to identify the importance of single climatic and hydrological events to sediment accumulation. Through automatic tomographic scanning of the sediment trap tube interior, we identified the daily changes of sediment accumulation rates with respect to meteorological events that were recorded at the meteorological station at the study site. The measured sediment volumes in the trap tube were very close to the manually measured sediment volumes (approximately 5% difference; Table 1), supporting the applicability of the tomographic scanning method for sediment flux rate studies. The deviation, however, increased up to 17% with very low sedimentation rates, which must be kept in mind if absolute values are compared. In most of the cases, tomographic scanning gave slightly larger values for sediment accumulation than manual measurements. However, even very low accumulation rates of less than 1 mL/day are recorded by tomographic scanning, and the theoretical accuracy of the method is ±0.00995 mL; the smallest measured value above zero was 0.00007 mL/day. The larger deviation of the October collection in the opposite direction compared to other measurements (Table 1) is very likely due to the sediment volume approaching the theoretical maximum. It seems that the practical maximum is slightly lower than the theoretical maximum. However, bearing these limitations in mind, we suggest that the trends, variability, and volumes of daily sediment accumulation are accurate enough to be used for monitoring of sediment flux variations. Large variability in the daily sediment flux data suggests the suitability of tomographic scanning for sediment flux measurements and underlines the need for ultra-high-resolution monitoring.

4.2. Sediment Flux Changes Controlled by Meteorological and Hydrological Events

Ultra-high-resolution observations and statistical analysis show a strong relationship between the rate of sediment accumulation in Savilahti Bay and meteorological and hydrological variables. The major parameters driving sediment accumulation on a seasonal scale are temperature, precipitation, and snow accumulation.

A positive correlation between sediment flux rate and temperature indicates spring, summer, and autumn as major seasons of sedimentation, which is in line with earlier sediment trap studies [14,15] as well as the conceptual models of varve formation in boreal environments [15,30,32]. A drop in sedimentation rate occurs for periods with temperatures below zero (Figure 5). This most likely indicates decreased sediment availability due to freezing and frost, as discussed earlier [4,14,21]. The nearly simultaneous ice cover (Figure 4) shows a negative correlation with sediment flux. Ice cover forms a physical barrier to allochthonous material transport and to wind and wave stress, as suggested earlier [23]. The importance of the ice cover period and frost formation for reduced sediment accumulation is supported by the *t*-test (Figure 8), which shows that the majority of sediment accumulation occurred during the ice- and frost-free period. However, during the winter (December–February), minor sediment deposition occurred under ice cover. These individual sedimentation events are associated with periods of increased air temperature (Figure 5).

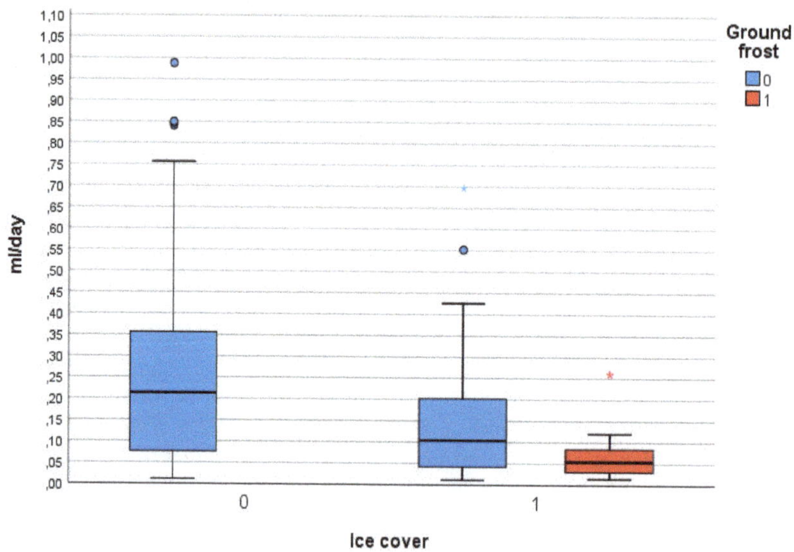

Figure 8. Boxplot of sediment flux rate with ice cover and ground frost grouping variables. Group value 0 means no ice cover and 1 means the presence of ice cover; for ground frost, group value 0 means no ground frost and 1 means the presence of ground frost at least in 1 cm deep. Note that the ground frost group value 1 boxplot is missing ice cover group 0, because there were no data for this group. Points represent outliers and asterisks represent extreme outliers.

A negative correlation coefficient between sediment flux rate and snow depth reflects the importance of snowmelt or loss for increased sediment accumulation, as suggested earlier [2,9,15,32]. In spring (March–May), the sedimentation flux rate started to increase on 12 April. This peak was due to snowmelt-induced spring flooding. The peak sediment accumulation rates between 12 and 19 April occurred at a time of rapid snow cover thickness loss (Figure 4). It is noteworthy how short and punctuated the first period of extreme sediment volumes is in spring. The snow around the lake melted before the ice cover melted and caused solids to move under the ice. Ground frost started to decrease on 6 March and reached 0 cm on 16 April, almost simultaneously with ice cover melt. However, after the ice and soil melted, the sedimentation rate increased again from late May to early June. This could be related to spring algal blooms [14,15,17]; however, the tomography method is incapable of defining the type of accumulating material.

A positive correlation between sediment flux rate and water temperature is related to spring, summer, and autumn with unfrozen soil, but this flux rate is also very likely influenced by autochthonous production and enhanced accumulation of biogenic matter [2,14,15,17]. Tomographic scanning cannot identify the type of sedimentary material, and thus the accumulation of organic matter compared to minerogenic matter cannot be discussed based on these results.

Precipitation in general has well-documented links of sediment accumulation in boreal zones [15,23,33]. Interestingly, during summer (June–August), intense water precipitation events caused increased sediment influxes with durations of several days (Figure 6). The precipitation-induced erosion along with the increased stream velocity and surface run-off likely explain the positive correlations. The lag could be related to the catchment response linked to, e.g., hydrological connectivity and soil retention, but also to varying settling velocity of particles with different size and density. This highlights the importance of not only winter precipitation and consequent spring flood, but also precipitation during the growing season, to sediment accumulation.

A positive significant correlation with wind speed reflects wind speed increases, leading to an increase in sediment flux rate. Enhanced wind speed during summer months is often related to storms and high precipitation rates and therefore is difficult to separate from the dominating influence of precipitation. When air and water temperatures are positive, wind-induced waves are likely to cause resuspension of littoral sediments, as suggested earlier [23].

During autumn (September–November), in October, the sedimentation rate decreased and air temperature fell to negative values, the topsoil froze, and solids were no longer prone to erosion due to precipitation. Topsoil started to freeze on 27 October and short and long precipitation periods occurred on 5–19 November.

Unfortunately, massive construction work started on the immediate catchment of the Savilahti Bay on 27 July 2018, and the sediment flux response to autumn conditions cannot be discussed with respect to climatic and hydrologic data, but was most likely driven by ditching and digging. This also led to unpredicted filling of the trap tube close to the theoretical maximum limit, leading to failure in the measurement of total volume by tomography compared to manual measurement (Table 1). However, the online trapping technique seems to be a very useful and valuable instrument in sediment flux studies to facilitate our understanding of climate-catchment processes, and also for measuring and monitoring the influence of anthropogenic activities on lake catchments. Online sediment trapping enables the investigation of the influence of single storm events, wind speed, and frost formation on sediment accumulation, which are discussed when interpreting varve records [2,4,15,21,23]. Such events are suggested to influence the sediment supply and measured lamina thicknesses, but so far this has been difficult to address. Combining the detailed analyses of sediment components and annual varve records could lead to an even deeper understanding of the driving mechanisms of sediment accumulation and lamina thickness. Further studies utilizing the long-term online trapping technique from basins with annually laminated sediments and open data sources, such as meteorological and hydrological observations, and their combination with data and statistical analysis methods will deepen our understanding of catchment dynamics and thus improve future interpretation based on high-resolution sediment sequences such as varved sediment records and sediment trap datasets.

5. Conclusions

Ultra-high-resolution sedimentation flux rates combined with meteorological and hydrological observations provide valuable information about the response of catchments to weather conditions and facilitates the interpretation of varve records. Daily measurements allow for a better understanding of the influence of single meteorological events such as heavy rainfall or increased wind speed on sediment transport and accumulation. Using annually laminated sediments and sediment trapping with deployment periods of several weeks, it is difficult to address these short-term events in detail. The ultra-high-resolution sediment flux time series sheds light on catchment responses to seasonal conditions and single meteorological events. Statistical analysis and time series disclose seasonal effects on sedimentation in the semiclosed Savilahti Bay basin in eastern Finland. Evidence on the importance of precipitation and temperature for sediment deposition is gained. Moreover, increased sediment fluxes lasting for several days following rainstorm events and enhanced wind speed are measured. The online trapping technique seems to be a very useful and valuable instrument for sediment flux studies. Further studies utilizing a long-term online trap technique and open data sources, such as meteorological and hydrological observations, and their combination with data and statistical analysis methods are suggested in order to deepen our understanding of catchment dynamics. This will facilitate paleoclimatic interpretations based on high-resolution sediment sequences such as varved sediment records.

Author Contributions: Conceptualization, M.J. and S.S.; methodology, M.J.; software, M.J.; validation, M.J., S.S., and J.S.; formal analysis, S.S.; investigation, M.J.; writing—original draft preparation, M.J.; writing—review and editing, S.S. and J.S.; visualization, M.J.; supervision, S.S.; project administration, J.S.

Funding: This research received no external funding.

Acknowledgments: We would like to sincerely thank junior researcher Samuel Hartikainen, University of Eastern Finland, for help installing trapping equipment in the field.

Conflicts of Interest: The authors declare no conflict of interest.

References

1. Ojala, A.E.K.; Francus, P.; Zolitschka, B.; Besonen, M.; Lamoreux, S.F. Characteristics of sedimentary varve chronologies—A review. *Quat. Sci. Rev.* **2012**, *43*, 45–60. [CrossRef]
2. Zolitschka, B.; Francus, P.; Ojala, A.E.K.; Schimmelmann, A. Varves in lake sediments—A review. *Quat. Sci. Rev.* **2015**, *117*, 1–41. [CrossRef]
3. Schimmelmann, A.; Lange, C.; Schieber, J.; Francus, P.; Ojala, A.E.K.; Zolitschka, B. Varves in marine sediments: A review. *Earth Sci. Rev.* **2016**, *159*, 215–246. [CrossRef]
4. Saarni, S.; Lensu, A.; Tammelin, M.; Haltia, E.; Saarinen, T. Winter climate signal in boreal clastic-biogenic varves; a comprehensive analysis of three varved records from 1890 to 1990 AD with meteorological and hydrological data from eastern Finland. *GFF* **2017**, *139*, 314–326. [CrossRef]
5. Amann, B.; Szidat, S.; Grosjean, M. A millennial-long record of warm season precipitation and flood frequency for the North-western Alps inferred from varved lake sediments: Implications for the future. *Quat. Sci. Rev.* **2015**, *115*, 89–100. [CrossRef]
6. Blass, A.; Bigler, C.; Grosjean, M.; Sturm, M. Decadal-scale autumn temperature reconstruction back to AD 1580 inferred from the varved sediments of Lake Silvaplana (southern Swiss Alps). *Quat. Res.* **2007**, *68*, 184–195. [CrossRef]
7. Moore, J.J.; Hughen, K.A.; Miller, G.H.; Overpeck, J.T. Little Ice Age recorded in summer temperature reconstructions from varved sediments of Donard Lake, Baffin Island, Canada. *J. Paleolimnol.* **2001**, *25*, 503–517. [CrossRef]
8. Brauer, A.; Dulski, P.; Mangili, C.; Mingram, J.; Liu, J. The potential of varves in high-resolution paleolimnological studies. *Pages News* **2009**, *17*, 96–98. [CrossRef]

9. Ojala, A.E.K.; Launonen, I.; Holmström, L.; Tiljander, M. Effects of solar forcing and North Atlantic Oscillation on the climate of continental Scandinavia during the Holocene. *Quat. Sci. Rev.* **2015**, *112*, 153–171.
10. Ólafsdóttir, K.B.; Geirsdóttir, Á.; Miller, G.H.; Larsen, D.J. Evolution of NAO and AMO strength and cyclicity derived from a 3-ka varve-thickness record from Iceland. *Quat. Sci. Rev.* **2013**, *69*, 142–154. [CrossRef]
11. Brauer, A.; Haug, G.H.; Dulski, P.; Sigman, D.M.; Negendank, J.F.W. An abrupt wind shift in western Europe at the onset of Younger Dryas cold Period. *Nat. Geosci.* **2008**, *1*, 520–523. [CrossRef]
12. Schletter, G.; Liu, Q.; Mingram, J.; Stebich, M.; Dulski, P. East-Asian monsoon variability between 15 000 and 2000 cal. yr BP recorded in varved sediments of Lake Sihailongwan (northeastern China, Long Gang volcanic field). *Holocene* **2006**, *16*, 1043–1057. [CrossRef]
13. Landmann, G.; Reimer, A.; Lemcke, G.; Kempe, S. Dating Late Glacial abrupt climate changes in the 14,570 yr long continuous varve record of Lake Van, Turkey. *Palaeogeogr. Palaeoclimatol. Palaeoecol.* **1996**, *122*, 107–118. [CrossRef]
14. Bonk, A.; Tylmann, W.; Amann, B.; Enters, D.; Grosjean, M. Modern limnology, sediment accumulation and varve formation processes in Lake Zabinskie, northeastern Poland: Comprehensive process studies as a key to understand the sediment record. *J. Limnol.* **2015**, *74*, 358–370. [CrossRef]
15. Ojala, A.E.K.; Kosonen, E.; Wecström, J.; Korkonen, S.; Korhola, A. Seasonal formation of clastic-boigenic varves: Potential for paleoenvironmental interpretations. *GFF* **2013**, *135*, 237–247. [CrossRef]
16. Trapote, M.C.; Vegas-Vilarrúbia, T.; López, P.; Puche, E.; Gomà, J.; Buchaca, T.; Cañellas-Boltà, N.; Safont, E.; Corella, J.P.; Rull, V. Modern sedimentary analogues and integrated monitoring to understand varve formation in the Mediterranean Lake Montcortès (Central Pyrenees, Spain). *Palaeogeogr. Palaeoclimatol. Palaeoecol.* **2018**, *496*, 292–304. [CrossRef]
17. Kaal, J.; Cortizas, A.M.; Rydberg, J.; Bigler, C. Seasonal changes in molecular composition of organic matter in lake sediment trap material from Nylandssjön, Sweden. *Org. Geochem.* **2015**, *83–84*, 253–262. [CrossRef]
18. Chu, G.; Liu, J.; Schettler, G.; Li, J.; Sun, Q.; Gu, Z.; Lu, H.; Liu, Q.; Liu, T. Sediment Fluxes and Varve Formation in Sihailongwan, a Maar Lake from Northeastern China. *J. Paleolimnol.* **2015**, *34*, 311–324. [CrossRef]
19. Tylmann, W.; Szpakowska, K.; Ohlendorf, C.; Woszczyk, M.; Zolitschka, B. Conditions for deposition of annually laminated sediments in a small meromictic lakes: A case study of Lake Suminko (northern Poland). *J. Paleolimnol.* **2012**, *47*, 55–70. [CrossRef]
20. Kienel, U.; Kirillin, G.; Brademann, B.; Plessen, B.; Lampe, R.; Brauer, A. Effects of spring warming and mixing duration on diatom deposition in deep Tiefer See, NE Germany. *J. Paleolimnol.* **2017**, *57*, 37–49. [CrossRef]
21. Saarni, S.; Muschitiello, F.; Weege, S.; Brauer, A.; Saarinen, T. A late Holocene record of solar-forced atmospheric blocking variability over Northern Europe inferred from varved lake sediments of Lake Kuninkaisenlampi. *Quat. Sci. Rev.* **2016**, *154*, 100–110. [CrossRef]
22. Martin-Puertas, C.; Brauer, A.; Dulski, P.; Brademann, B. Testing climate-proxy stationarity throughout the Holocene: An example from the varved sediments of Lake Meerfelder Maar (Germany). *Quat. Sci. Rev.* **2012**, *58*, 56–65. [CrossRef]
23. Itkonen, A.; Salonen, V.-P. The response of sedimentation in three varved lacustrine sequences to air temperature, precipitation and human impact. *J. Paleolimnol.* **1994**, *11*, 323–332. [CrossRef]
24. The Finnish Meteorological Institute's Open Data. Available online: https://en.ilmatieteenlaitos.fi/open-data (accessed on 30 March 2019).
25. Finland's Environmental Administration, Hydrological Observations. Available online: https://www.ymparisto.fi/en-US/Maps_and_statistics/Hydrological_observations (accessed on 30 March 2019).
26. Saarnisto, M. Shoreline displacement and emergence of lake basins. Geological Survey of Finland. *Spec. Pap.* **2000**, *29*, 25–34.
27. Eronen, M.; Haila, H. Geologinen kehitys jääkauden lopussa. In *Atlas of Finland, Surficial Deposits*; Alalammi, P., Ed.; National Board of Survey and Geographical Society of Finland: Helsinki, Finland, 1990; Volume 124, pp. 13–19. (In Finnish)
28. Johansson, M.; Saarni, S.; Sorvari, J. A modern on-line technology for environmental monitoring: Tomographic approach in sediment trap provides record of sediment flux with ultra-high-resolution. Forthcoming.
29. IBM Corp. *IBM SPSS Statistics for Windows, Version 25.0*; IBM Corp.: Armonk, NY, USA, 2017.

30. Haltia-Hovi, E.; Saarinen, T.; Kukkonen, M. A 2000-year record of solar forcing on varved lake sediments in eastern Finland. *Quat. Sci. Rev.* **2007**, *26*, 678–689. [CrossRef]
31. Czymzik, M.; Brauer, A.; Dulski, P.; Plessen, B.; Naumann, R.; von Grafesnstein, U.; Scheffler, R. Orbital and solar forcing of shifts in Mid- to Late Holocene flood intensity from varved sediments of pre-Alpine Lake Ammersee (southern Germany). *Quat. Sci. Rev.* **2015**, *61*, 96–110. [CrossRef]
32. Renberg, I. Formation, structure and visual appearance of iron-rich, varved lake sediments. *Proc. Int. Assoc. Theor. Appl. Limnol.* **1981**, *21*, 1922–2010. [CrossRef]
33. Saarni, S.; Saarinen, T.; Lensu, A. Organic lacustrine sediment varves as indicators of past precipitation chamges: A 3000-year climate record from Central Finland. *J. Paleolimnol.* **1995**, *53*, 401–413. [CrossRef]

quaternary

MDPI

Article

Varve Distribution Reveals Spatiotemporal Hypolimnetic Hypoxia Oscillations During the Past 200 Years in Lake Lehmilampi, Eastern Finland

Sarianna Salminen [1,*], Saija Saarni [1], Mira Tammelin [1], Yu Fukumoto [2] and Timo Saarinen [1]

1 Department of Geography and Geology, University of Turku, FIN-20014 Turku, Finland; saitur@utu.fi (S.S.); mira.tammelin@utu.fi (M.T.); tijusa@utu.fi (T.S.)

2 Research center for Pan-Pacific civilizations, Ritsumeikan University, Kyoto 603–8577, Japan; misudotomakudo@yahoo.co.jp

* Correspondence: sarsalm@utu.fi; Tel.: +358-408-484-783

Received: 15 January 2019; Accepted: 22 May 2019; Published: 26 May 2019

Abstract: We investigated 34 sediment cores to reconstruct spatiotemporal variations in hypolimnetic hypoxia for the past 200 years in Lehmilampi, a small lake in Eastern Finland. As hypoxia is essential for varve preservation, spatiotemporal changes in varve distribution were used as an indicator for hypolimnetic hypoxia oscillations. The hypoxic water volume was used as a variable reflecting hypolimnetic hypoxia and determined for each year by estimating the water volume beneath the water depth where shallowest varves were preserved. As a result, seven hypoxia periods, highlighting the variations in hypolimnetic hypoxia, are established. These periods may be influenced by bioturbation, lake infill, and lake level changes. Furthermore, we evaluated the relationship between hypolimnetic hypoxia oscillations and climatic factors. Diatom assemblage changes were also analyzed to estimate whether the hypoxia periods could be related to anthropogenic eutrophication. The diatom analyses suggest relatively stable nutrient conditions for the past 200 years in Lake Lehmilampi. Climate, on the other hand, seems to be an important driver of hypoxia oscillations based on correlation analysis. The role of individual forcing factors and their interaction with hypolimnetic hypoxia would benefit from further investigations. Understanding climatic and anthropogenic forcing behind hypolimnetic hypoxia oscillations is essential when assessing the fate of boreal lakes in a multi-stressor world.

Keywords: varves; hypoxia; oxygen deficiency; lake sediments; eutrophication; hypolimnetic hypoxia oscillations

1. Introduction

Eutrophication and hypoxia have become environmental challenges worldwide, causing loss of biodiversity, fish kills, and algal blooms [1–3]. Therefore, they are also an economic problem [4]. Eutrophication in lakes is triggered by excess nutrients [5] leading to hypolimnetic hypoxia through enhanced autochthonous production and increased oxygen consumption caused by the degradation of biogenic material [6,7]. The nutrient overload is generally a result of anthropogenic influence, such as sewage and runoff from agricultural fields, construction areas, and industry [8]. In addition to anthropogenic factors, changes in climate affect hypolimnetic oxygen concentrations. For instance, the solubility of oxygen in water corresponds negatively to increasing water temperatures [9]. Increasing water temperature may also reduce the mixing of oxygen-rich surface waters with deeper waters [9] and thus extend the stratification period [10]. Hence, variations in climate trigger variations in hypolimnetic hypoxia, i.e., their depth, area, and volume. These variations in hypoxia are hereinafter called hypolimnetic hypoxia oscillations.

Lake sediments are natural archives suitable for investigating past climatic and anthropogenic changes in the environment [11–13]. Their formation is mainly controlled by climate, catchment conditions (e.g., weathering, surface run-off, groundwater flow), and anthropogenic factors [14,15]. Varves are annual laminations that represent seasonal sedimentation cycles in a lake [12,16] and can be considered as indicators of hypolimnetic hypoxia [17,18]. High lacustrine productivity and seasonal variability in runoff transporting eroded clastic material into a lake are essential for varve formation [12,19]. Varve preservation, however, requires certain features [20,21] that support the absence of sediment mixing, erosion, and re-suspension. Hypoxia (≤ 2 mg L^{-1}) [22], the oxygen level under which freshwater organisms cannot survive [1], is the key to further the preservation of varves [21,23] because it results in the absence of bioturbation. It more likely occurs in stratified and eutrophic lakes with oxygen consumption due to organic matter decay (mineralization) [7]. In contrast to longer-term hypoxia, seasonal hypoxia might not result in varve formation, for instance, because of uniform seasonal sedimentation or wind-induced resuspension [24].

Although the approach of using the presence of varves to study hypolimnetic hypoxia has shown to be suitable, the number of studies using varves to establish the onset of hypoxia oscillations is limited [17,18,25]. The present study details spatiotemporal changes in varve preservation from the deepest point towards the shore of Lake Lehmilampi in Eastern Finland during the past 200 years. As varve preservation is highly dependent on hypoxia, the study also relates to hypolimnetic hypoxia oscillations. In addition, we evaluated possible forcing factors (climate variations and eutrophication) influencing hypolimnetic hypoxia oscillations in Lake Lehmilampi.

2. Study Site

Lake Lehmilampi (Figure 1a,b) is a small headwater lake in Eastern Finland (63°37′42 N, 29°06′09 E, surface area 0.17 km^2) with a northern (maximum water depth: 10.8 m) and a southern basin (maximum water depth: 11.6 m). Both basins are elongated in the northwest-southeast direction. The northern shoreline of the northern basin and the southern shoreline of the southern basin are rather steep. Lake Lehmilampi has two inflows, one from the southeast and another from the southwest and one outflow to the northwest. Lake water monitoring data is available only for three dates since the 1970s showing seasonal and annual variations in water quality (Figure 1c).

Hydrological monitoring data are unavailable for this small lake but exist for the larger and deeper (maximum water depth: 61 m) neighboring Lake Pielinen, located ca. 9 km to the south from Lake Lehmilampi. The freezing date varies from October 23rd to January 29th (average November 25th) in Lake Pielinen, melting date from April 29th to June 3rd (average May 18th), and ice cover duration from 105–212 days (average 175 days) [26,27]. Spring and autumn overturns typically take place in May and September–October, respectively [28]. The water level in Lake Pielinen (Figure 1d) decreases to its minimum in April, after which it increases to its maximum in June–July [29]. The average annual difference between the minimum and maximum water levels is 118 cm [29].

Lake Lehmilampi is located in the boreal climate zone. Mean annual temperature is 3.2 °C, mean spring (March, April, May) temperature 2.2 °C, mean summer (June, July, August) temperature 15.2 °C, mean autumn (September, October, November) temperature 3.7 °C, and mean winter (December, January, February) temperature −8.5 °C at the weather station Kuopio Maaninka [30]. This station is located approximately 100 km southwest from Lake Lehmilampi. Mean annual precipitation is 612 mm, with approximately half as rainfall and the rest as snowfall [28]. Mean spring precipitation is 112 mm, mean summer precipitation 218 mm, mean autumn precipitation 158 mm, and mean winter precipitation 124 mm. In Eastern Finland, wind velocities are rather low, and mean annual wind speed varies between 2.5–3.4 m/s at Lake Lehmilampi [31]. Wind velocities are highest during autumn and lowest during winter. Generally, the wind direction is southeast during autumn and southwest during the rest of the year.

Land use in Eastern Finland changed into a more permanent slash-and-burn cultivation in the 16th century [32,33], although paleolimnological analyses suggest sporadic, small-scale land use already during prehistoric times, i.e., prior to the 14th century [34,35]. Cultivation intensified in the 18th century in the vicinity of Lake Lehmilampi, as evidenced by increasing cereal pollen in the lake sediments [35]. Traditional slash-and-burn agriculture continued in Eastern Finland until the late 19th century [36]. After the Second World War, agriculture, forestry, and urbanization intensified in Finland [37]. Draining peatlands for forestry was most extensive in Eastern Finland during the 1960s–1980s [38], and a small drained peatland also exists in the Lake Lehmilampi catchment. Currently, the catchment is mainly forested (Figure 1e). The bedrock in the catchment area (1 km^2) is composed of Archean tonalite (a plutonic rock compositionally between granite and diorite), whereas the surface geology (Figure 1f) is dominated by Quaternary till and silty loams [39].

The sediments at the two deepest basins of Lake Lehmilampi are varved, representing seasonal changes in sediment accumulation [40]. The basic varve structure is composed of clastic and biogenic laminae (Figure 2). The clasts are eroded from the catchment and transported into the lake by spring floods following snowmelts. As a consequence, a sharp contact is typical between the dark-colored organic lamina and the following light-colored clastic lamina, whereas the contact is gradual between the clastic lamina and the following biogenic lamina [40]. The biogenic lamina is composed of autochthonous and allochthonous organic matter deposited during the growing season from summer to autumn [40,41]. The thickness of biogenic lamina depends on the primary production rate, transport of autochthonous biogenic material, preservation, and degradation [42]. Preservation and degradation are closely linked to hypolimnetic oxygen concentrations [43,44]. According to an earlier study [45], varve preservation in the northern basin of Lake Lehmilampi started approximately 5100 BP when the lake became isolated from Lake Pielinen.

3. Materials and Methods

3.1. Coring and Sample Preparation

A comprehensive sediment sampling of Lake Lehmilampi was performed in March 2014 when the lake was frozen. Altogether, 14 long piston cores (diameter 5 cm, length 140–167 cm) and 20 short freeze cores were recovered along two transects (Figure 1a,b, Table 1) at water depths between 6.5–10.8 m, to obtain a detailed perception of sediment characteristics throughout the lake. Piston cores were recovered with a lightweight Livingstone piston corer and stored in a cold room at 5 °C. They were spliced into two halves along their long axis with a circular saw and a wire. Sediment surfaces were cleaned with a glass slip and covered with a thin plastic film for magnetic susceptibility measurements. The freeze cores were recovered with an HTH sediment corer [46] applying the ice-finger technique [47,48] with dry ice to obtain a frozen surface sample. The freeze cores were approximately 30 cm long and represent ca. 200 years at the most. These cores were stored in a freezer at −18 °C. To preserve varve structures, the freeze cores were impregnated with epoxy resin using the water-acetone-epoxy exchange method [49,50]. The cured samples were sawn using a rock saw and polished into 1.8 mm thin slabs for incident light microscopy analysis and X-ray radiography. To perceive the basic Lake Lehmilampi varve structure in detail, a 30 μm thin section was prepared using an ASTERA CUT8 diamond saw (Astera Solutions, Zürich, Switzerland, manufactured in Vaasa, Finland) and an ASTERA GRN16 grinder (Astera Solutions, Zürich, Switzerland, manufactured in Vaasa, Finland).

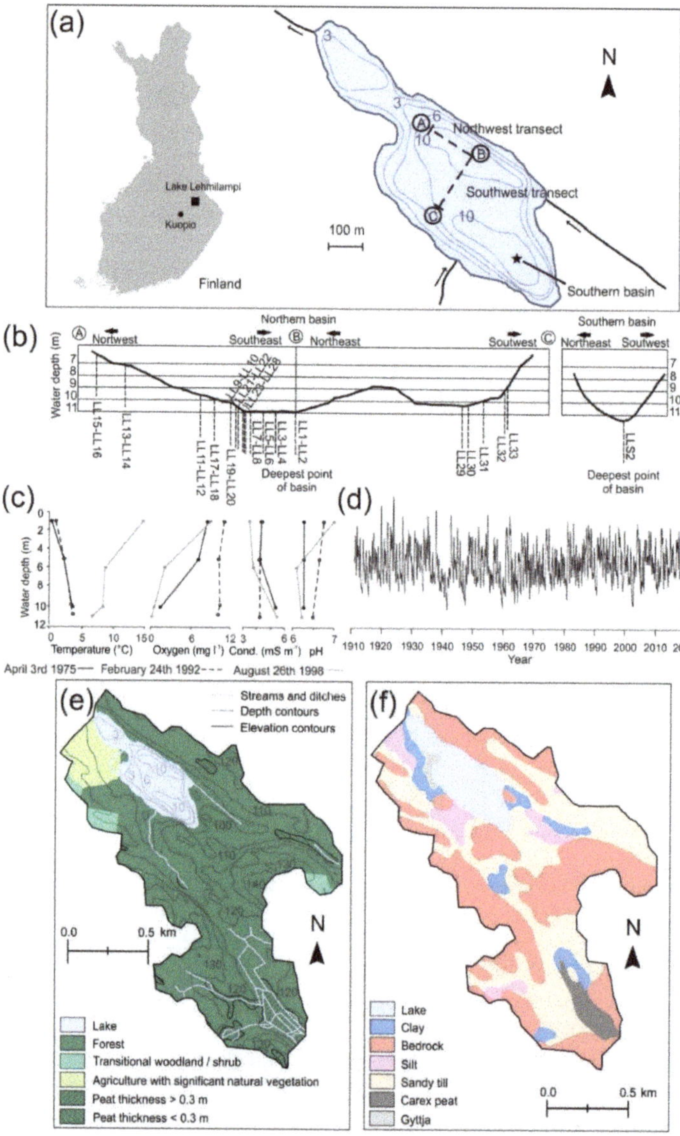

Figure 1. Lake location, sampling transects, water quality and water level data, and drainage basin: (**a**) Location of and sampling transects through Lake Lehmilampi. Dashed lines show the northwest (AB) and the southwest transects (BC). The star denotes the sampling point in the southern basin. (**b**) Coring sites and core codes along both transects are shown as a function of bathymetry. (**c**) Water quality monitoring data of Lake Lehmilampi (Cond.: electrical conductivity). The lake has been frozen in April 3rd, 1975 and in February 24th, 1992. The lake has been free of ice in August 26th, 1998. (**d**) Water level data of neighboring Lake Pielinen. Lake levels are expressed as height (cm) above an arbitrary zero level at the monitoring location. (**e**) Land use types of the Lake Lehmilampi drainage basin [51–53]. Isobaths are expressed in meter below lake surface, and elevation contour values are expressed as meters above sea level. (**f**) Superficial geology of the Lake Lehmilampi drainage basin [51,52,54].

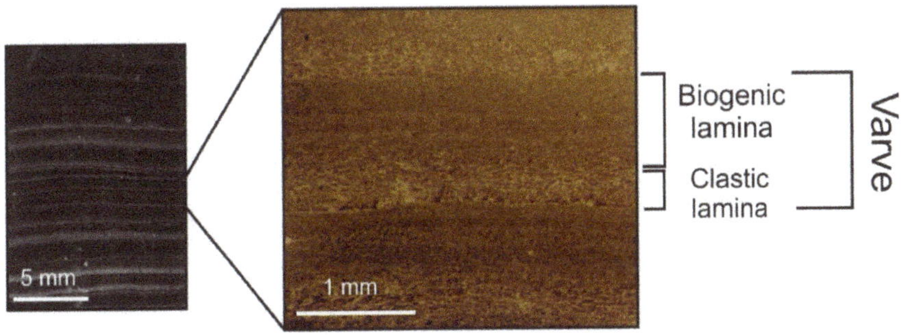

Figure 2. Photography of polished freeze core surface (left) and enlarged thin section microscopy image of the varve structure (right) from Lake Lehmilampi representing the seasonal sedimentation cycle. Light-colored clastic laminae are composed of clastic material transported into the lake during spring snowmelt flooding. Darker-colored biogenic laminae are composed of organic matter produced during the growing season.

Table 1. Piston cores (PC) and freeze cores (FC) recovered from Lake Lehmilampi (AB: northwest transect, BC: southwest transect, NB: northern basin, SB: southern basin). Sediment types are characterized as varved (a), partially varved (b), and non-varved (c).

Core	Core Type	Sampling Point/ Transect	Water Depth (m)	Sediment Types	Core Length (cm)	Core	Core Type	Sampling Point/ Transect	Water Depth (m)	Sediment Types	Core Length (cm)
LL1	FC	NB	10.80	a	32	LL18	PC	AB transect	9.90	a,b,c	140
LL2	PC	NB	10.80	a	142	LL19	FC	AB transect	10.30	a,b,c	29
LL3	FC	AB transect	10.80	a	26	LL20	PC	AB transect	10.30	a,b,c	152
LL4	PC	AB transect	10.80	a	147	LL21	FC	AB transect	10.60	a,b,c	32
LL5	FC	AB transect	10.80	a	30	LL22	PC	AB transect	10.60	a,b,c	140
LL6	PC	AB transect	10.80	a	152	LL23	FC	AB transect	10.75	a	27
LL7	FC	AB transect	10.80	a	33	LL24	PC	AB transect	10.75	a	142
LL8	PC	AB transect	10.80	a	162	LL25	FC	AB transect	10.75	a	37
LL9	FC	AB transect	10.50	a,b	21	LL26	PC	AB transect	10.75	a	166
LL10	PC	AB transect	10.50	a,b	167	LL27	FC	AB transect	10.75	a	32
LL11	FC	AB transect	9.58	a,b,c	17	LL28	PC	AB transect	10.75	a	146
LL12	PC	AB transect	9.58	a,b,c	149	LL29	FC	BC transect	10.40	a,b	27
LL13	FC	AB transect	7.20	a,b,c	24	LL30	FC	BC transect	10.30	a,b,c	33
LL14	PC	AB transect	7.20	a,b,c	158	LL31	FC	BC transect	9.90	a,b,c	24
LL15	FC	AB transect	6.53	c	25	LL32	FC	BC transect	9.10	a,b,c	30
LL16	PC	AB transect	6.53	c	162	LL33	FC	BC transect	8.68	a,b,c	28
LL17	FC	AB transect	9.90	a,b,c	27	LLS2	FC	SB	11.60	a	32

3.2. *Varve Analysis*

Varves were counted from each epoxy-impregnated and polished freeze core sub-sample with the help of incident light microscopy and X-ray images. Incident light microscopy images were taken with a Canon EOS 600D and Canon EOS Utility software coupled with a Nikon SMZ800 stereomicroscope. X-ray radiography was performed at the University of Helsinki using the μCT scanner Nanotom 180 NF. To establish spatiotemporal changes in varve distribution, the sediment was divided into different sections by varve analysis. For each freeze core, varved (a), partially varved (b), and non-varved sections (c) were identified (Figure 3, Table 1). Varve, clastic lamina and biogenic lamina thicknesses were measured from the epoxy-impregnated sub-samples using a microscope to study variations in sediment composition. For each freeze core, the percentage of varved sediments between the years 1815 and 2014 was calculated by measuring the thickness of varved sections and comparing it to the entire core length. For each non-varved section, sediment accumulation rates were calculated as an average of the section in question.

Sections are considered as varved (Figure 3a) if:

1. Varves are present in incident light microscopy images or X-ray images.

2. Varves exhibit both a clastic spring lamina and a biogenic growing season lamina with clear differences in color or brightness.
3. Varves have a sharp contact between preceding organic and subsequent clastic lamina.

Sections are considered as partially varved (Figure 3b) if:

1. Varve structure without a sharp contact between preceding organic and subsequent clastic lamina is identified.
2. Laminae are disturbed or laterally discontinuous. Such sections sometimes consist of several subsequent varves.

Sections are considered as non-varved (Figure 3c) if:

1. Sediment is massive
2. No laminated structure is identified.

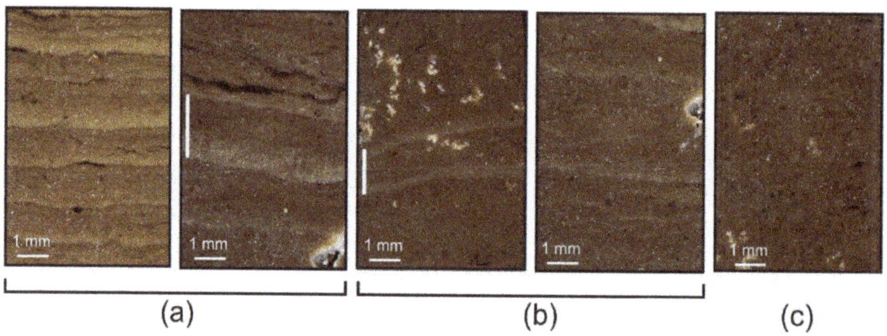

Figure 3. Examples of different sediment types in incident light microscopy images. (**a**) Varved sediment: continuous varved section (left) and one clastic-organic couplet (right; marked with a white bar). (**b**) Partially varved sediment with unclear contacts and disturbed or discontinuous lamina. White bar (left) represents partially varved sediment comprising two clastic laminae and one biogenic lamina. (**c**) Non-varved massive section.

3.3. *Dating and Core Correlation*

Core LL1 from the deepest point of the northern basin was dated using varve counting and considered as the master core, because of its undisturbed and continuous varve record. A similar record has also been observed in earlier studies of the northern basin [40,45]. Radiometric dating was performed for core LL1 to verify the accuracy of varve counting. ^{137}Cs activities were measured with a BrightSpec gamma spectrometer on wet sediment samples of 11–30 g with 3600 s counting time at the Geological Survey of Finland. Each wet sediment sample integrated two varve years.

All cores, particularly the varved sections, were correlated with core LL1 using seasonal changes in varve succession, lithological marker horizons (e.g., particularly thick clastic laminae) that represent major changes in sediment composition, as well as magnetic susceptibility. Varve succession and lithological markers were identified from incident light microscopy and X-ray images, the latter showing density variations of the sediment (Figure 4). The beginning and the end of the non-varved and partially varved sections were correlated with the master core based on the varves below and above the non-varved and partially varved sections. Magnetic susceptibility measurements were carried out for each piston core using the Bartington MS2 susceptibility meter connected to a Bartington MS2E core logging sensor at 2 mm intervals on an automatic measuring track.

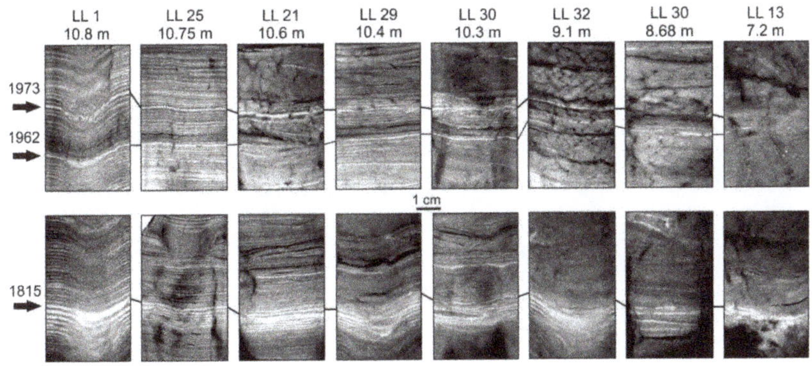

Figure 4. X-ray images representing core correlation based on variations in lithological marker horizons and the succession of varves. The arrows point at clastic laminae that were formed in 1815, 1962, and 1973 as easily discernible components of marker horizons.

3.4. Modeling the Hypoxic Water Volume

Hypoxia throughout the lake was modeled using the distribution of varves during the past 200 years (1815–2014). The annual spatial distribution of varves in the lake was determined based on the water depth of the varve preserving sediments. The determined water depth represents the boundary below which varves have been preserved. For each year, the spatial distribution of the varved area and the water volume below the critical water depth corresponding to this area were determined with the ArcGIS ArcMap 10.1 software. Modeling of the hypoxic water volume was based on the assumption that the volume of Lake Lehmilampi has remained relatively constant during the study period. Possible effects of lake level changes and lake infill are discussed in Section 5. The bathymetric curves of Lake Lehmilampi were interpolated with the nearest neighbor interpolation method using measurement data from the National Land Survey of Finland (NLS) data service [55] together with water depth measurements taken during sampling.

We refer to the percentage of hypoxic water volume of the whole lake water volume as the hypoxia volume. The water volume data were divided into two groups, hypoxia and baseline periods, based on the median hypoxia volume, which was defined as the cutoff value for baseline conditions of hypoxia in Lake Lehmilampi. Annual hypoxia volumes below this cutoff value were considered as baseline periods, whereas hypoxia volumes at or exceeding the cutoff value were classified as hypoxia periods.

3.5. Meteorological and Diatom Data

To identify possible environmental factors triggering spatiotemporal changes in hypoxia in Lake Lehmilampi, we acquired meteorological data from NORDKLIM (Dataset for Climate Analysis with Data from the Nordic Region) [56] and analyzed the diatom assemblages of each hypoxia and baseline period from core LL1. The long meterorological data of Kuopio Maaninka weather monitoring station covers monthly data for 1890–1998. We calculated annual temperature and precipitation, seasonal temperature and precipitation, and the number of days with snow cover (>50% covered by snow). Moreover, monthly spring (March, April, May) temperature and precipitation as well as monthly autumn (September, October, November) temperature and precipitation were calculated to investigate possible effects of spring and autumn overturns.

Diatoms were analyzed because they are sensitive indicators of environmental change and widely used for reconstructing past nutrient concentrations of lakes [57]. Each diatom sample represents one hypoxia or baseline period, apart from the longest hypoxia periods 1 (3 samples) and 7 (2 samples), and the shortest hypoxia period 5 (no samples). The diatom samples integrate 3–12 years, depending on the duration of the period in question. One varve at the start and one at the end of each period were excluded to avoid contamination between subsequent periods. Diatom slides were prepared using standard procedures [57]. At least 300 diatom valves were identified to species level under an optical microscope at x1000 magnification using the taxonomical references of Krammer and Lange-Bertalot [58–61] and updated nomenclature of Porter [62], Spaulding et al. [63], and Guiry and Guiry [64]. Higher taxonomic level summaries were made by calculating relative abundances of planktic large-celled Aulacoseira taxa, small-celled cyclotelloid taxa, and elongate pennate taxa, as well as small, benthic fragilarioid taxa. The diatom stratigraphy was plotted using TILIA and TILIAGRAPH software [65].

3.6. Statistical Analyses

Correlations of hypoxia volumes with varve parameters and meteorological data were calculated using the software R3.4.3 [66] to identify the most significant covariates. As the volume of hypoxic water was not normally distributed, Spearman's correlation analysis [67] was applied. In order to take decadal trends into account, 10-year moving averages were calculated from annual, seasonal, and monthly meteorological data. The diatom assemblage changes were investigated with principal component analysis (PCA) [68] using Canoco 4.5 for Windows [69]. Spearman's correlations were calculated between the sample scores of the first (PC1) and second (PC2) principal components, the summarized diatom groups, and diatom-inferred total phosphorus (DI-TP). DI-TP was reconstructed for each diatom sample using the diatom-TP transfer function of Tammelin et al. [70] and C2 software version 1.6.8 [71] to estimate possible changes in the nutrient status of Lake Lehmilampi. The transfer function is based on central-eastern Finnish lakes and weighted averaging partial least squares regression (WA-PLS) [72]. Its jack-knifed coefficient of determination (r^2_{jack}) is 0.72, and its root mean squared error of prediction ($RMSEP_{jack}$) is 0.191 log $\mu g\ L^{-1}$ [70].

4. Results

4.1. Varve Characteristics

The topmost 3.2 cm sediment of core LL1, representing 15 years (1999–2014), is rather unconsolidated compared to the sediment below. Varve, biogenic lamina and clastic lamina thicknesses decreased correspondingly downcore, except for the high clastic lamina thickness at the beginning of the studied time period (Figure 5a). Clastic/biogenic lamina thickness ratio does not show a clear trend, but instead, its highest values occurred at both ends of the observed time period. The average sedimentation rate is 1.21 mm/year (range 0.4–4.8 mm/year) at the deepest point of the northern basin and decreases along the transects towards shallower depths. At a water depth of 9.6 m, it is 0.65 mm/year in the BC transect and 0.8 mm/year in the AB transect. The varve percentages decrease in relation to water depth from 100% at the deepest point of the basin to 0% at depths shallower than 7.2 m (Table 2, Figure 5b). Furthermore, magnetic susceptibility shows comparable variations at different water depths (Figure 5c).

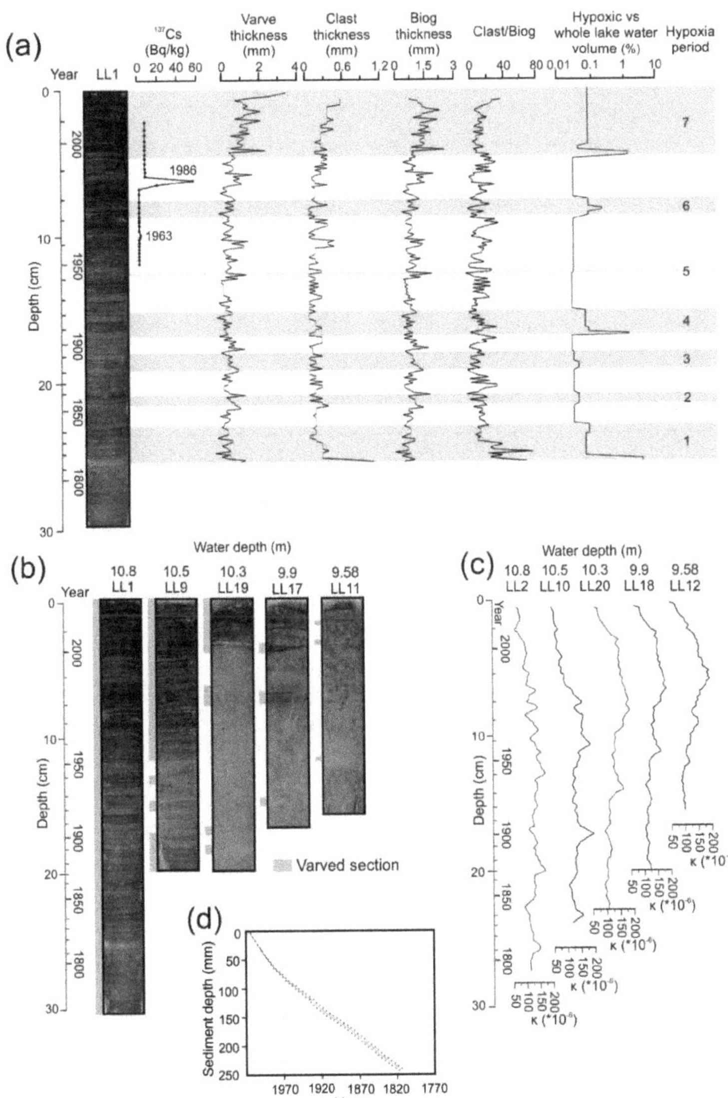

Figure 5. Sediment characteristics and dating: (**a**) Photography of core LL1 (water depth: 10.8 m), ^{137}Cs activity measured from LL1 and variations in varve thickness, clastic (Clast) and biogenic (Biog) lamina thicknesses, as well as clastic/biogenic lamina thickness (Clast/Biog) ratio. Hypoxia volume from 1815–2014 and hypoxia periods (gray bars) are also provided. Hypoxia volume is displayed on a logarithmic scale. (**b**) Example of sediment appearance in freeze cores along the AB transect of Lake Lehmilampi and (**c**) example of variations in magnetic susceptibility (κ). Core codes and water depths are provided. Varved sections are marked with gray bars to the left of the photographs. Note that the depth scale is valid for all cores, while the time scale represents only the master core LL1. (**d**) Age versus sediment depth curve with estimated varve counting error (dashed lines) for core LL1.

Table 2. Percentages of varved sediments for freeze cores from different water depths.

Core	Water Depth (m)	Varve Percentage (%)
LL1-LL7	10.80	100
LL23-LL27	10.75	100
LL21	10.60	69
LL9	10.50	68
LL29	10.40	35
LL19	10.30	27
LL30	10.30	22
LL17	9.90	12
LL31	9.90	11
LL11	9.58	8
LL32	9.10	6
LL33	8.68	5
LL13	7.20	1
LL15	6.53	0

4.2. Dating and Core Correlation

According to the varve chronology, core LL1 from the deepest point of the northern basin represents approximately 200 years of sedimentation (Figure 5a,d). Cumulative varve counting error is ±0.5% (199 ± 1 years) (Figure 5d). Two ^{137}Cs peaks were present in core LL1. The major Cs activity peak in Figure 5a represents the Chernobyl accident in 1986. The second, minor peak is rather weak, but possibly represents atmospheric nuclear tests in 1963. The peaks are in agreement with the varve counts of continuously varved core LL1. Therefore, we assume that its varve chronology is reliable and hypoxic conditions and varve preservation have been continuous at the deepest point of the basin during the observed period. Varve preservation in relation to water depth appeared similar in both transects, and the succession of varves and lithological marker horizons are traceable between cores allowing their correlation (Figure 4). Similarities in the magnetic susceptibility profiles of the piston cores further support our core correlation (Figure 5c).

4.3. Changes in Varve Distribution and Hypoxic Water Volume

The area and volume of hypoxic waters fluctuated in Lake Lehmilampi during the past 200 years according to varve-based modeling. Years 1815, 1855, 1975, and 2014 are portrayed in Figure 6 because they represent the most and least extensive hypoxia periods displaying the variations in hypoxia. The magnitude of hypoxia volume ranged between 0.03–5.7%. The median hypoxia volume of the whole data, i.e., the cutoff value for the baseline conditions of hypoxia, was 0.046%. Consequently, seven hypoxia periods were identified from 1815–2014 (Table 3, Figure 5a, and Figure 7). These periods were rather short-lived compared to the time between them, apart from the longer hypoxia periods 1, 4, and 7 (Figure 5a). Although short-lived, the small increase in hypoxia volume during the 1940s was considered as hypoxia period 5.

Table 3. Hypoxia periods, their durations, median hypoxia volume (m^3), range of variation of hypoxia volume (m^3), and median and maximum hypoxia volume per lake volume (%).

Hypoxia Period	Duration (Years)	Duration (from-to)	Median Hypoxia Volume (m^3)	Range of Variation of Hypoxia Volume (m^3)	Median Hypoxia Volume per Lake Volume (%)	Maximum Hypoxia Volume per Lake Volume (%)
7	20	1994–2014	2255	787–34599	0.240	3.727
6	9	1971–1980	1638	1638–7648	0.180	0.823
5	1	1943	428	428–428	0.046	0.046
4	16	1905–1921	1638	1638–34599	0.180	3.727
3	11	1881–1892	428	428–787	0.046	0.085
2	5	1852–1857	787	787–787	0.084	0.085
1	27	1815–1842	1638	1638–53304	0.180	5.740

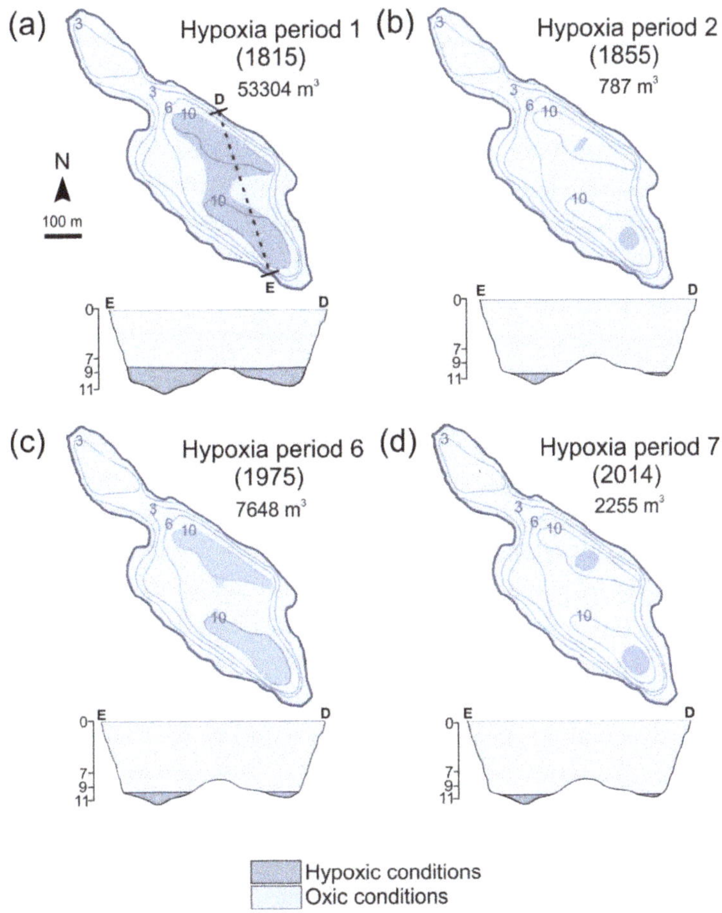

Figure 6. First year of occurrence and volume of hypoxic waters (in m^3) for (**a**) hypoxia period 1 in 1815, (**b**) hypoxia period 2 in 1855, (**c**) hypoxia period 6 in 1975, and (**d**) hypoxia period 7 in 2014. Bathymetric cross-sections along the dashed line DE show hypoxic (darker color) and oxic conditions (lighter color). Water depths 7–11 m in the transects are exaggerated by a factor of 25 to better visualize the variations in hypolimnetic hypoxia.

4.4. Climate and Varve Thickness Correlations with Hypoxia Volume

Hypoxia volumes have statistically significant (p-value < 0.05) correlation coefficients (ρ) higher than 0.3 with six climate variables (Table 4, Figure 7). Of these six variables, winter and March temperatures correlate positively with hypoxia, whereas the correlation is negative between hypoxia and the number of days with snow cover, November temperature, as well as autumn and October precipitation. Winter temperature ($\rho = 0.56$, p-value < 0.01) and October precipitation ($\rho = -0.56$, p-value < 0.01) have highest correlation coefficients. Clastic lamina thickness and hypoxic water volume showed a weak positive correlation ($\rho = 0.23$, p-value < 0.01), whereas total varve thickness and biogenic lamina thickness have no statistically significant correlations with hypoxia (p-value > 0.05).

Table 4. Correlation between hypoxia volume and studied climate variables. From the monthly spring and autumn temperature and precipitation variables, only those with statistically significant correlations (p-value < 0.05) are shown. Time periods of available monitoring data, number of data points (n), correlation coefficients (ρ), and p-values are provided. * Variables with correlation coefficients > 0.3 and p-values < 0.05.

Variable	Period	n	ρ	p-Value
Annual temperature	1890–1997	107	0.36	0.06
Annual precipitation	1890–1997	107	−0.05	0.63
Winter temperature *	1890–1997	107	0.56	<0.01
Winter precipitation	1890–1997	107	−0.06	0.22
Days of snow cover *	1957–1997	40	−0.38	0.03
Spring temperature	1890–1997	107	0.27	<0.01
March temperature *	1890–1997	107	0.39	<0.01
Spring precipitation	1890–1997	107	−0.11	0.27
Summer temperature	1890–1997	107	−0.26	0.01
Summer precipitation	1890–1997	107	0.26	<0.01
Autumn temperature	1890–1997	107	−0.23	0.02
November temperature *	1890–1997	107	−0.47	<0.01
Autumn precipitation *	1890–1997	107	−0.36	<0.01
October precipitation *	1890–1997	107	−0.56	<0.01

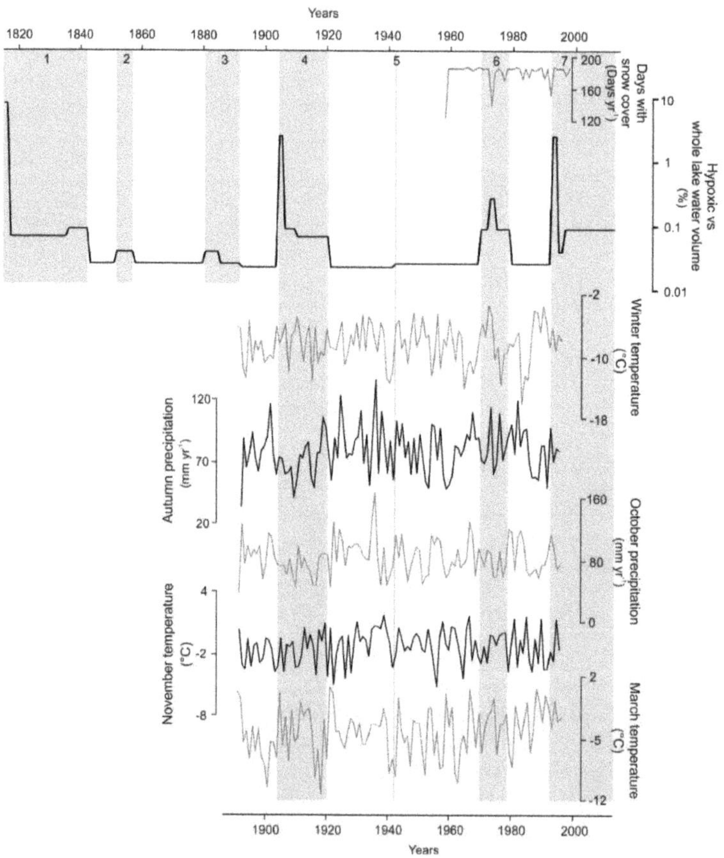

Figure 7. Hypoxic water volume from 1815–2014 compared to variations in climate variables. Hypoxia volume is displayed on a logarithmic scale. Climate variables with correlation coefficients > 0.3 and p-values < 0.05 are shown. Hypoxia periods 1–7 are marked with grey bars.

4.5. Diatom Analysis

The diatom assemblages of core LL1 changed during the past 200 years, while DI-TP remained relatively stable (Figure 8). The diatom assemblage change is characterized by an increase in the relative abundance of small cyclotelloid taxa and elongate pennate taxa and a decrease in large *Aulacoseira* taxa and small fragilarioid taxa towards the present day. PC1, i.e., largest variation in the diatom data, correlates negatively with large *Aulacoseira* taxa ($\rho = -0.93$, p-value < 0.01) and benthic fragilarioid taxa ($\rho = -0.55$, p-value $= 0.04$). Small cyclotelloid taxa have a positive correlation with PC1 ($\rho = 0.80$, p-value < 0.01). PC2, i.e., second largest variation in diatom data, correlates positively with elongate pennates ($\rho = 0.66$, p-value $= 0.01$). DI-TP does not correlate (p-value > 0.05) with PC1 or PC2.

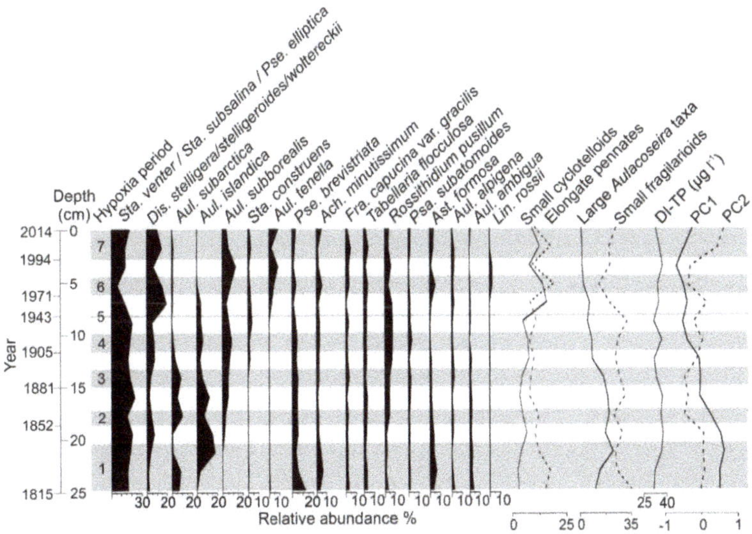

Figure 8. Relative abundances of most abundant diatom species and four higher-level taxonomic groups of core LL1. The sample scores of the first two principal components (PC1 and PC2) of diatom data and diatom-inferred total phosphorus (DI-TP), and temporal coverage of diatom samples are also shown. Hypoxia periods 1–7 are marked with grey bars. Years shown along the age/depth scale mark the onset of hypoxia periods.

5. Discussion

5.1. Spatiotemporal Changes in Varves and Hypolimnetic Hypoxia

The global spread of hypoxia in lakes started in the latter half of the 19th century and accelerated after the Second World War, particularly in densely populated regions [73]. Our results from the remote boreal Lake Lehmilampi, however, suggest spatiotemporal fluctuations in varve preservation and hypoxia instead of an increasing trend towards the present day (Figure 5a, Table 3). Oxygen-depleted conditions have prevailed for thousands of years in the northern basin of Lake Lehmilampi based on its continuous varve preservation [45]. The boundary between continuous and discontinuous varve preservation (and hypoxia) is located relatively close to the maximum depths of the basins at water depths of 10.60–10.75 m, whereas the boundary between discontinuous and no varve preservation lies at 6.53–7.20 m. The limited water quality data from Lake Lehmilampi (Figure 1c) suggests that hypoxia may temporarily reach even slightly shallower depths. Certain characteristics of Lake Lehmilampi, such as the strong seasonal contrasts in boreal climate, elongated and small but relatively deep basins, and higher-relief bedrock outcrops sheltering the lake from winds, are favorable for the formation and preservation of varves [19].

We identified seven hypoxia periods for Lake Lehmilampi, from which the first (1815–1842) was the longest and most intensive (Figure 6, Figure 7, Table 3). It presumably started earlier than 1815 and existed for more than 27 years. Longer sediment cores would be needed to confirm these assumptions. The onset of a hypoxia period appears to be rapid and intensive (average duration: 1.5 years), whereas the return back to baseline conditions seems less pronounced (average duration: 8.1 years). These observations are in accordance with the previously reported slow responses in hypolimnetic hypoxia to reduced external forcing [73]. Apart from highest peak values during the onset of the three longest hypoxia periods (periods 1, 4, and 7), the overall hypoxic water volume has remained relatively low and hypoxic bottom water conditions have been restricted to a small area compared to the lake surface area (Figures 5a and 6, Table 3). The three most extensive peaks in hypoxic volume (3.7–5.7% of the whole lake water volume), however, are comparable to the 8% increase in hypoxic volume of Lake Bourget in the French Alps (CE 1930–1960) resulting from a complex interaction of anthropogenic and natural forcing factors [24].

Bioturbation and lake infill may have led to an underestimation of some of the hypoxia periods. The effect of lake infill increases downcore because hypoxia volume and lake volume were slightly higher when less sediment had been accumulated to the deep basins. Bioturbation, on the other hand, could disturb previously preserved varve structures [1,74]. Moreover, the prolonged return from hypoxia to baseline conditions could be a result of the gradual return of bioturbation. Abundance and distribution of lacustrine bioturbation are not yet well known and vary within as well as between lakes [75]. The penetration depth of benthic fauna depends on oxygenation, sedimentation rate, and grain size [76,77]. Shallow-burrowing animals are often the most abundant and have patchy distributions [75]. Better knowledge of lacustrine bioturbation would be beneficial for reconstructing lake histories [75], including past changes in hypolimnetic hypoxia.

5.2. Potential Forcing Factors behind Hypolimnetic Hypoxia Oscillations

Despite anthropogenic eutrophication often being the main driver of hypoxia in European lakes [24,70], it does not seem to be a major forcing factor in Lake Lehmilampi during the past 200 years, although humans were present in the vicinity of the lake already prior to our study interval [35]. Disentangling the effects of eutrophication and climate change can be difficult [78], but diatom assemblages suggest relatively stable nutrient conditions in Lake Lehmilampi and are more likely responding to climate-induced changes in the water column (Figure 8). The main variation in the diatom data of Lake Lehmilampi (PC1), characterized by a recent increase in small cyclotelloid taxa relative to large *Aulacoseira* taxa and small fragilarioids, is typical to lakes experiencing increased thermal stability and longer periods of thermal stratification due to anthropogenic climate warming [79]. This main variation seems unrelated to hypoxia events. Nevertheless, the relative abundance of elongate pennates increases particularly during hypoxia events 1, 6, and 7, which could be a response to multiple stressors, such as climate warming, eutrophication, or increasing dissolved organic matter [79,80]. *Aulacoseira subarctica* (Müller) Haworth, a taxon favoring moderately nutrient-enriched and sufficiently turbulent lakes [81], also appears to be more abundant during hypoxia periods 1, 2, and 3. *Aulacoseira islandica* (Müller) Simonsen, on the contrary, does not show a similar resemblance with the hypoxia periods, although its autecological preferences are similar to *A. subarctica* [82].

Strong and unambiguous relationships between hypoxia volume and varve, clastic lamina, or biogenic lamina thicknesses were not found in Lake Lehmilampi. For instance, hypoxia period 1 coincided with increased clastic lamina thickness, and hypoxia period 7 had a simultaneous increase in biogenic lamina thickness, whereas hypoxia period 4 did not seem to be clearly related to increases in varve or lamina thicknesses at all. The positive correlation between clastic lamina thickness and hypoxia volume implies that hypoxia could be associated with catchment erosion, as erosion typically influences clastic lamina accumulation via spring floods [83–85]. Nevertheless, the correlation is weak, suggesting partially different or more complex interactions between the forcing factors behind hypoxia and varve/lamina thicknesses. Augustsson et al. [35] linked the highest erosion intensities of Lake

Lehmilampi to periods of local fire and human impact. Haltia-Hovi et al. [37] found cyclic variation in varve thickness and composition of Lake Lehmilampi during the last 2000 years and interpreted it as reflecting changes in climate, likely driven by solar forcing.

The climate seems to be an important driver of changes in Lake Lehmilampi during the last 200 years despite human influence, because the recent diatom assemblage change is likely related to a warming climate, and several statistically significant correlations were found between climate variables and hypoxia volume (Table 4). These correlations reflect relationships between yearly, seasonal, and monthly variations and do not take into account single extreme events or possible time lags between climate forcing and responses in hypolimnetic hypoxia. Nevertheless, increased hypoxia volumes seem to coincide at least with milder winters and drier Octobers (Table 4). Climate controls the alternating stratification and overturn periods in boreal lakes [86], which also affect oxygen availability. Therefore, it is not unexpected to find correlations between climate variables and hypoxia volumes. Dry Octobers could reduce the intensity of autumn overturn via diminished runoff of oxygen-rich water into the lake. However, the positive correlation between hypoxia volume and winter temperature is interesting and unexpected, because cold winters typically extend ice cover, which weakens circulation in the lake through thermal stratification and reduces oxygen availability in bottom waters [16]. The mechanisms beyond these correlations cannot be explained using our data and remain as topics for further investigations.

Water level monitoring data is not available for the small and remote Lake Lehmilampi. Thus, its effect on hypolimnetic hypoxia oscillations could not be estimated and leaves the following discussion on a speculative level. Over a century-long water level monitoring data of the notably larger Lake Pielinen (into which Lake Lehmilampi flows through Lake Roukkajanjärvi and the River Ylikylänjoki) shows seasonal changes in water level but not a long-term trend (Figure 1d). The water level of Lake Lehmilampi could be sensitive to changes in precipitation because the lake is large in relation to its catchment area [87]. In humid climates, a decrease in precipitation is more likely to influence lake levels than an increase in precipitation because the latter leads to increased outflow compensating increased runoff [88]. Lake Lehmilampi is a headwater lake, so it is not influenced by upstream lakes or rivers. Groundwater flow is probably negligible because aquifers in Finland are commonly located in areas of coarse-grained sandy and gravelly deposits [86], which do not occur in the catchment of Lake Lehmilampi [41,54].

Understanding fully the variations in varve distribution and hypolimnetic hypoxia oscillations requires evaluation of several natural and anthropogenic factors for pre-anthropogenic and anthropogenic time periods. Lake Lehmilampi provides an excellent varve record spanning thousands of years, but monitoring data is unfortunately limited, typical for small and remote boreal lakes. Detailed future studies between relationships of lake level variations and hypoxia volumes with distribution and intensity of bioturbation in boreal lakes would greatly improve our understanding of hypolimnetic hypoxia oscillations. Also extending hypoxia volume modeling all the way back to the isolation of Lake Lehmilampi or other varved boreal lakes would give interesting information on the long-term natural and anthropogenic changes in hypolimnetic hypoxia.

6. Conclusions

Our study of 34 sediment cores along two transects in varved, boreal Lake Lehmilampi shows spatiotemporal variations in varve preservation during the past 200 years. As varve preservation is closely linked with oxygen-depleted conditions in the bottom waters, we use spatiotemporal variations in varve preservation as a proxy for hypolimnetic hypoxia oscillations. Seven hypoxia periods, exceeding the low baseline value for hypoxic water volume in Lake Lehmilampi, were established based on changes in varve distribution. This hypoxia modeling was based on the assumption of relatively constant lake water volume during the past 200 years, and the influence of potential lake level changes, bioturbation, and lake infill are discussed on a speculative level. Examination of possible environmental drivers behind these hypoxia periods suggested that hypolimnetic hypoxia oscillations in Lake Lehmilampi are more likely related to climatic than anthropogenic forcing. Diatom assemblage changes indicate relatively stable nutrient conditions and several climate variables correlate with hypolimnetic hypoxia during the past 200 years. These climate variables could have influenced hypolimnetic hypoxia, for instance, through changes in the intensity of catchment erosion and water mixing during overturns. Understanding the role of individual forcing factors and their complex interaction with varve preservation and long-term hypolimnetic hypoxia oscillations would benefit from further investigations.

Author Contributions: Conceptualization, S.S. (Sarianna Salminen) and S.S. (Saija Saarni); Methodology, S.S. (Sarianna Salminen) and M.T.; Software, S.S. (Sarianna Salminen), M.T.; Validation, S.S. (Sarianna Salminen); Formal Analysis, S.S. (Sarianna Salminen), M.T. and Y.F.; Investigation, S.S. (Sarianna Salminen); Resources, T.S.; Data Curation, S.S. (Sarianna Salminen); Writing—Original Draft Preparation, S.S. (Sarianna Salminen); Writing—Review & Editing, S.S. (Saija Saarni), M.T., T.S., and Y.F.; Visualization, S.S. (Sarianna Salminen); Supervision, T.S.; Project Administration, S.S. (Sarianna Salminen); Funding Acquisition, S.S. (Sarianna Salminen), T.S.

Funding: This study was funded by Maa- ja vesitekniikan tuki ry. and The Academy of Finland (269834).

Acknowledgments: We thank H. Wenho, A. Peltola, and K. Kashima for technical support in the field, M. Kalliokoski for the assistance in GIS analyses, and J. Manninen for thin section preparation. We thank E. Haltia, anonymous reviewers, and the Guest Editor for valuable comments on the manuscript.

Conflicts of Interest: The authors declare no conflict of interest. The sponsors had no role in the design, execution, interpretation, or writing of the study.

References

1. Diaz, R.J. Overview of hypoxia around the world. *J. Environ. Qual.* **2001**, *30*, 275–281. [CrossRef]
2. Baulch, H.M.; Schindler, D.W.; Turner, M.A.; Findlay, D.L.; Paterson, M.J.; Vinebrook, R.D. Effects of warming on benthic communities in a boreal lake: Implications of climate change. *Limnol. Oceanogr.* **2005**, *50*, 1377–1392. [CrossRef]
3. Koutsodendris, A.; Brauer, A.; Zacharias, I.; Putyrskaya, V.; Klemt, E.; Sangiorgi, F.; Pross, J. Ecosystem response to human-and climate-induced environmental stress on an anoxic coastal lagoon (Etoliko, Greece) since 1930 AD. *J. Paleolimnol.* **2015**, *53*, 255–270. [CrossRef]
4. Wilson, M.A.; Carpenter, R. Economic valuation of freshwater ecosystem services in The United States: 1971–1997. *Ecol. Appl.* **1999**, *9*, 772–783.
5. Schindler, D.W. Evolution in Phosphorus Limitation in Lakes. *Science* **1977**, *195*, 260–262. [CrossRef]
6. Müller, B.; Bryant, L.D.; Matzinger, A.; Wüest, A. Hypolimnetic oxygen depletion in eutrophic lakes. *Environ. Sci. Technol.* **2012**, *46*, 9964–9971. [CrossRef] [PubMed]
7. Friedrich, J.; Janssen, F.; Aleynik, D.; Bange, H.W.; Boltacheva, N.; Çagatay, M.N.; Dale, A.W.; Etiope, G.; Erdem, Z.; Geraga, M.; et al. Investigating hypoxia in aquatic environments: Diverse approaches to addressing a complex phenomenon. *Biogeosciences* **2014**, *11*, 1215–1259. [CrossRef]
8. Carpenter, S.R. Eutrophication of aquatic ecosystems: Bistability and soil phosphorus. *Proc. Natl. Acad. Sci. USA* **2005**, *102*, 10002–10005. [CrossRef] [PubMed]
9. Deutsch, C.; Brix, H.; Ito, T.; Frenzel, H.; Thompson, L. Climate-Forced Variability of Ocean Hypoxia. *Science* **2011**, *333*, 336. [CrossRef]
10. Livingstone, D.M. Impact of secular climate change on the thermal structure of a large temperate central European lake. *Clim. Chang.* **2003**, *57*, 205–225. [CrossRef]
11. Last, W.M.; Smol, J.P. An introduction to physical and geochemical methods used in paleolimnology. In *Tracking Environmental Change Using Lake Sediments. Vol. 1, Basin Analysis, Coring, and Chronological Techniques*; Last, W.M., Smol, J.P., Eds.; Kluwer Academic Publishers: Dordrecht, The Netherlands, 2001; pp. 1–5.
12. Zolitschka, B.; Francus, P.; Ojala, A.E.; Schimmelmann, A. Varves in lake sediments—A review. *Quat. Sci. Rev.* **2015**, *117*, 1–41. [CrossRef]
13. Zolitschka, B.; Enters, D. Lacustrine sediments. In *Encyclopedia of Paleoclimatology and Ancient Environments*; Gornitz, V., Ed.; Springer: Dordrecht, The Netherlands, 2009; pp. 485–488.

14. Ohlendorf, C.; Sturm, M.; Hausmann, S. Natural environmental changes and human impact reflected in sediments of a high alpine lake in Switzerland. *J. Paleolimnol.* **2003**, *30*, 297–306. [CrossRef]
15. Tyllmann, W. Lithological and geochemical record of anthropogenic changes in recent sediments of a small and shallow lake (Lake Pusty Staw, northern Poland). *J. Paleolimnol.* **2005**, *33*, 313–325. [CrossRef]
16. O'sullivan, P.E. Annually-laminated lake sediments and the study of Quaternary environmental changes—A review. *Quat. Sci. Rev.* **1983**, *1*, 245–313. [CrossRef]
17. Giguet-Covex, C.; Arnaud, F.; Poulenard, J.; Enters, D.; Reyss, J.L.; Millet, L.; Lazzaroto, J.; Vidal, O. Sedimentological and geochemical records of past trophic state and hypolimnetic anoxia in large, hard-water Lake Bourget, French Alps. *J. Paleolimnol.* **2010**, *43*, 171–190. [CrossRef]
18. Jenny, J.P.; Arnaud, F.; Dorioz, J.M.; Giguet Covex, C.; Frossard, V.; Sabatier, P.; Millet, L.; Reyss, J.L.; Tachikawa, K.; Bard, E.; et al. A spatiotemporal investigation of varved sediments highlights the dynamics of hypolimnetic hypoxia in a large hard-water lake over the last 150 years. *Limnol. Oceanogr.* **2013**, *58*, 1395–1408. [CrossRef]
19. Ojala, A.E.K.; Francus, P.; Zolitschka, B.; Besonen, M.; Lamoureux, S.F. Characteristics of sedimentary varve chronologies—A review. *Quat. Sci. Rev.* **2012**, *43*, 45–60. [CrossRef]
20. Tylmann, W.; Zolitschka, B.; Enters, D.; Ohlendorf, C. Laminated lake sediments in northeast Poland: Distribution, preconditions for formation and potential for paleoenvironmental investigation. *J. Paleolimnol.* **2013**, *50*, 487–503. [CrossRef]
21. Ojala, A.E.; Saarinen, T.; Salonen, V.P. Preconditions for the formation of annually laminated lake sediments in southern and central Finland. *Boreal Environ. Res.* **2000**, *5*, 243–255.
22. Roberts, J.J.; Höök, T.O.; Ludsin, S.A.; Pothoven, S.A.; Vanderploeg, H.A.; Brandt, S.B. Effects of hypolimnetic hypoxia on foraging and distributions of Lake Erie yellow perch. *J. Exp. Mar. Biol. Ecol.* **2009**, *381*, 132–142. [CrossRef]
23. Saarnisto, M. Annually laminated lake sediments. In *Handbook of Holocene Palaeoecology and Palaeohydrology*; Berglund, B.E., Ed.; John Wiley & Sons Ltd.: New Jersey, USA, 1986; pp. 343–370.
24. Jenny, J.P.; Normandeau, A.; Francus, P.; Taranu, Z.E.; Gregory-Eaves, I.; Lapointe, F.; Jautzy, J.; Ojala, A.E.K.; Dorioz, J.M.; Schimmelmann, A.; et al. Urban point sources of nutrients were the leading cause for the historical spread of hypoxia across European lakes. *Proc. Natl. Acad. Sci. USA* **2016**, *113*, 12655–12660. [CrossRef]
25. Kienel, U.; Dulski, P.; Ott, F.; Lorenz, S.; Brauer, A. Recently induced anoxia leading to the preservation of seasonal laminae in two NE-German lakes. *J. Paleolimnol.* **2013**, *50*, 535–544. [CrossRef]
26. Korhonen, J. *Suomen Vesistöjen Jääolot. Finnish Environment, 751*; Finnish Environment Institute Reports; Finnish Environment Institute: Helsinki, Finland, 2005; (In Finnish, Summary "Ice Conditions in Lakes and Rivers in Finland" in English); Available online: https://helda.helsinki.fi/handle/10138/40687 (accessed on 3 March 2019).
27. Korhonen, J. Long-term changes in lake ice cover in Finland. *Nord. Hydrol.* **2006**, *37*, 347–363. [CrossRef]
28. Hertta Database. Available online: https://www.syke.fi/en-US/Open_information (accessed on 15 March 2016).
29. Korhonen, J. *Suomen Vesistöjen Virtaaman ja Vedenkorkeuden Vaihtelut. Finnish Environment 45*; Finnish Environment Institute Reports; Finnish Environment Institute: Helsinki, Finland, 2007; Finnish Environment Institute Reports; (In Finnish, Summary "Discharge and water level variations in lakes and rivers in Finland" in English); Available online: https://helda.helsinki.fi/handle/10138/38428 (accessed on 3 March 2019).
30. Pirinen, P.; Simola, H.; Aalto, J.; Kaukoranta, J.P.; Karlsson, P.; Ruuhela, R. *Tilastoja Suomen Ilmastosta 1981–2010*; Finnish Environment Institute Reports; Finnish Environment Institute: Helsinki, Finland, 2012; Available online: https://helda.helsinki.fi/handle/10138/35880 (accessed on 7 March 2019).
31. Finnish Meteorological Institute. Weather Data Download Service. Available online: https://en.ilmatieteenlaitos.fi/download-observations#!/ (accessed on 1 April 2019).
32. Soininen, A. The Colonization of Northern Savo in the 15th and 16th Centuries. In *Historiallisia Tutkimuksia LVIII. Suomen Historiallinen Seura*; Finnish Environment Institute: Helsinki, Finland, 1961; (In Finnish, summary in English); Available online: https://www.sgr.fi/sust/sust265/sust265_korpela.pdf (accessed on 12 August 2018).

33. Korpela, J. Migratory Lapps and the population explosion of Eastern Finns: The early colonization of Eastern Finland reconsidered. In Proceedings of the Networks, Interaction and Emerging Identities in Fennoscandia and Beyond, Finno-Ugrian Society Conference, Tromsø, Norway, 13–16 October 2009.
34. Taavitsainen, J.P.; Simola, H.; Grönlund, E. Cultivation history beyond the periphery: Early agriculture in the North European boreal forest. *J. World Prehist.* **1998**, *12*, 199–253. [CrossRef]
35. Augustsson, A.; Gaillard, M.J.; Peltola, P.; Mazier, F.; Bergbäck, B.; Saarinen, T. Effects of land use and climate change on erosion intensity and sediment geochemistry at Lake Lehmilampi, Finland. *Holocene* **2013**, *23*, 1247–1259. [CrossRef]
36. Soininen, A. Old Traditional Agriculture in Finland in the 18th and 19th Centuries. In *Historiallisia Tutkimuksia 96. Suomen Historiallinen Seura*; Finnish Environment Institute: Helsinki, Finland, 1974; (In Finnish, summary in English); Available online: http://www.doria.fi/handle/10024/167610 (accessed on 24 May 2019).
37. Simola, H.; Arvola, L. Lakes of Northern Europe. In *The Lakes Handbook Volume 2: Lake Restoration and Rehabilitation*; O'Sullivan, P.E., Reynolds, C.S., Eds.; Blackwell Publishing Ltd.: Oxford, UK, 2005; pp. 117–158.
38. Turunen, J. Development of Finnish peatland area and carbon storage 1950–2000. *Boreal Environ. Res.* **2008**, *13*, 319–334.
39. Maankamara e DigiKP. Digital Map Database. Available online: http://gtkdata.gtk.fi/maankamara/ (accessed on 16 April 2017).
40. Haltia-Hovi, E.; Saarinen, T.; Kukkonen, M. A 2000-year record of solar forcing on varved lake sediment in eastern Finland. *Quat. Sci. Rev.* **2007**, *26*, 678–689. [CrossRef]
41. Meyers, P.A.; Lallier-Vergès, E. Lacustrine sedimentary organic matter records of Late Quaternary paleoclimates. *J. Paleolimnol.* **1999**, *21*, 345–372. [CrossRef]
42. den Heyer, C.; Kalff, J. Organic matter mineralization rates in sediments: A within—And among-lake study. *Limnol. Oceanogr.* **1998**, *43*, 695–705. [CrossRef]
43. Hedges, J.I.; Hu, F.S.; Devol, A.H.; Hartnett, H.E.; Tsamakis, E.; Keil, R.G. Sedimentary organic matter preservation; a test for selective degradation under oxic conditions. *Am. J. Sci.* **1999**, *299*, 529–555. [CrossRef]
44. Sobek, S.; Durisch-Kaiser, E.; Zurbrügg, R.; Wongfun, N.; Wessels, M.; Pasche, N.; Wehrli, B. Organic carbon burial efficiency in lake sediments controlled by oxygen exposure time and sediment source. *Limnol. Oceanogr.* **2009**, *54*, 2243–2254. [CrossRef]
45. Haltia-Hovi, E.; Nowaczyk, N.; Saarinen, T.; Plessen, B. Magnetic properties and environmental changes recorded in Lake Lehmilampi (Finland) during the Holocene. *J. Paleolimnol.* **2010**, *43*, 1–13. [CrossRef]
46. Renberg, I.; Hansson, H. The HTH sediment corer. *J. Paleolimnol.* **2008**, *40*, 655–659. [CrossRef]
47. Renberg, I. Improved methods for sampling, photographing and varve-counting of varved lake sediments. *Boreas* **1981**, *10*, 255–258. [CrossRef]
48. Saarinen, T.; Wenho, H. Minijääsormi sekä muita uusia ja vanhoja ideoita järvisedimentin talvikairaukseen. In Proceedings of the Congress Abstract Book, Geologian Tutkijapäivät, Turku, Finland, 14–15 March 2005; pp. 72–73.
49. Lamoureux, S.F. Embedding unfrozen lake sediments for thin section preparation. *J. Paleolimnol.* **1994**, *10*, 141–146. [CrossRef]
50. Tiljander, M.; Ojala, A.; Saarinen, T.; Snowball, I. Documentation of the physical properties of annually laminated (varved) sediments at a sub-annual to decadal resolution for environmental interpretation. *Quat. Int.* **2002**, *88*, 5–12. [CrossRef]
51. Finnish Environment Institute 2014. Downloadable Spatial Datasets. Catchment Areas. Available online: https://www.syke.fi/en-US/Open_information/Spatial_datasets (accessed on 21 March 2019).
52. National Land Survey of Finland 2017. Topographic Database. Available online: https://tiedostopalvelu.maanmittauslaitos.fi/tp/kartta?lang=en. (accessed on 21 March 2019).
53. Finnish Environment Institute 2018. Corine Land Cover. Available online: https://www.syke.fi/en-US/Open_information/Spatial_datasets. (accessed on 21 March 2019).
54. Geological Survey of Finland 2015. Superficial Deposits. Available online: https://hakku.gtk.fi/en/locations/search. (accessed on 4 December 2018).
55. National Land Survey of Finland. Paituli Spatial Data Download Service. Available online: www.csc.fi/paituli (accessed on 14 March 2016).
56. NORDKLIM Dataset (NKDS) v2 2015. Available online: http://blog.fmi.fi/nordmet/node/214 (accessed on 15 January 2017).

57. Battarbee, R.W.; Jones, V.J.; Flower, R.J.; Cameron, N.G.; Bennion, H.; Carvalho, L.; Juggins, S. Diatoms. In *Tracking Environmental Change Using Lake Sediments*; Smol, J.P., Birks, H.J.B., Last, W.M., Eds.; Kluwer Academic Publishers: Dordrecht, The Netherlands, 2001; Volume 3, pp. 155–202.
58. Krammer, K.; Lange-Bertalot, H. Bacillariophyceae, Teil: Naviculaceae. In *Süsswasserflora von Mitteleuropa*; Ettle, H., Gerloff, J., Heynig, H., Mollenhauer, D., Eds.; Fischer: New York, NY, USA, 1986; Band 2/1.
59. Krammer, K.; Lange-Bertalot, H. Bacillariophyceae, Teil: Bacillariaceae, Epithemiaceae, Surirellaceae. In *Süsswasserflora von Mitteleuropa*; Ettle, H., Gerloff, J., Heynig, H., Mollenhauer, D., Eds.; Fischer: Stuttgart, Germany; New York, NY, USA, 1988; Band 2/2.
60. Krammer, K.; Lange-Bertalot, H. Bacillariophyceae, Teil: Centrales, Fragilariaceae, Eunotiaceae. In *Süsswasserflora von Mitteleuropa*; Ettle, H., Gerloff, J., Heynig, H., Mollenhauer, D., Eds.; Fisher: Stuttgart, Germany; Jena, Germany, 1991; Band 2/3.
61. Krammer, K.; Lange-Bertalot, H. Bacillariophyceae, Teil: Achnanthaceae. In *Süsswasserflora von Mitteleuropa*; Ettle, H., Gartner, G., Gerloff, J., Heynig, H., Mollenhauer, D., Eds.; Fisher: Stuttgart, Germany; Jena, Germany, 1991; Band 2/4.
62. Porter, S.P. Algal Attributes: An Autecological Classification of Algal Taxa Collected by the National Water-Quality Assessment Program. In *US Geological Survey Data Series 329*; U.S. Geological Survey: Reston, VA, USA, 2008.
63. Spaulding, S.A.; Lubinski, D.J.; Potapova, M. Diatoms of the United States. 2008. Available online: http://westerndiatoms.colorado.edu. (accessed on 16 October 2018).
64. Guiry, M.D.; Guiry, G.M.; AlgaeBase. World-Wide Electronic Publication, National University of Ireland, Galway. 2018. Available online: http://www.algaebase.org (accessed on 24 October 2018).
65. Grimm, E. *TILIA and TILIAGRAPH Pollen Diagramming Program*; Illinois State Museum: Springfield, IL, USA, 1991.
66. R Development Core Team. *R: A Language and Environment for Statistical Computing*; R Foundation for Statistical Computing: Vienna, Austria, 2016.
67. Spearman, C. The proof and measurement of association between two things. *Am. J. Psychol.* **1904**, *15*, 72–101. [CrossRef]
68. Pearson, K. On lines and planes of closest fit to systems of points in space. *Philos. Mag.* **1901**, *2*, 559–572. [CrossRef]
69. Ter Braak, C.J.F.; Šmilauer, P. *CANOCO Reference Manual and Cano Draw for Windows User's Guide: Software for Canonical Community Ordination*; Version 4.5; Microcomputer Power: Ithaca, NY, USA, 2002.
70. Tammelin, M.; Kauppila, T.; Viitasalo, M. Factors controlling recent diatom assemblages across a steep local nutrient gradient in central-eastern Finland. *Hydrobiologia* **2017**, *799*, 309–325. [CrossRef]
71. Juggins, S. *C2 Version 1.5 User Guide, Software for Ecological and Palaeoecological Data Analysis and Visualization*; University of Newcastle: Newcastle Upon Tyne, UK, 2007.
72. Ter Braak, C.J.F.; Juggins, S. Weighted averaging partial least squares regression (WA-PLS): An improved method for reconstructing environmental variables from species assemblages. *Hydrobiologia* **1993**, *269*, 485–502. [CrossRef]
73. Jenny, J.P.; Francus, P.; Normandeau, A.; Lapointe, F.; Perga, M.E.; Ojala, A.; Schimmelmann, A.; Zolitschka, B. Global spread of hypoxia in freshwater ecosystems during the last three centuries is caused by rising local human pressure. *Glob. Chan. Biol.* **2016**, *22*, 1481–1489. [CrossRef]
74. Xie, M.; Wang, N.; Gaillard, J.F.; Packman, A.I. Interplay between flow and bioturbation enhances metal efflux from low-permeability sediments. *J. Hazard. Mater.* **2018**, *341*, 304–312. [CrossRef] [PubMed]
75. White, D.S.; Miller, M.F. Benthic invertebrate activity in lakes: Linking present and historical bioturbation patterns. *Aquat. Biol.* **2008**, *2*, 269–277. [CrossRef]
76. Savrda, C.E. Taphonomy of trace fossils. In *Trace Fossils: Concepts, Problems, Prospects*; Miller, W., III, Ed.; Elsevier: Oxford, UK, 2007; pp. 92–109.
77. Wetzel, A. Ecologic interpretation of deep-sea trace fossil communities. *Palaeogeogr. Palaeoclimatol. Palaeoecol.* **1991**, *85*, 47–69. [CrossRef]
78. Davidson, T.A.; Jeppesen, E. The role of palaeolimnology in assessing eutrophication and its impact on lakes. *J. Paleolimnol.* **2013**, *49*, 391–410. [CrossRef]
79. Rühland, K.M.; Paterson, A.M.; Smol, J.P. Lake diatom responses to warming: Reviewing the evidence. *J. Paleolimnol.* **2015**, *54*, 1–35. [CrossRef]

80. Kissman, C.E.H.; Williamson, C.E.; Rose, K.C.; Saros, J.E. Response of phytoplankton in an alpine lake to inputs of dissolved organic matter through nutrient enrichment and trophic forcing. *Limnol. Oceanogr.* **2013**, *58*, 867–880. [CrossRef]
81. Gibson, C.E.; Anderson, N.J.; Haworth, E. Aulacoseira subarctica: Taxonomy, physiology, ecology and palaeoecology. *Eur. J. Phycol.* **2003**, *38*, 83–101. [CrossRef]
82. Stoermer, E.F.; Kreis, R.G., Jr.; Sicko-Goad, L. A Systematic, Quantitative, and Ecological Comparison of Melosira Islandica, O. Müll. with M. Granulata (EHR.) Ralfs from the Laurentian Great Lakes. *J. Great Lakes Res.* **1981**, *7*, 345–356. [CrossRef]
83. Ojala, A.E.; Alenius, T. 10 000 years of interannual sedimentation recorded in the Lake Nautajärvi (Finland) clastic–organic varves. *Palaeogeogr. Palaeoclimatol. Palaeoecol.* **2005**, *219*, 285–302. [CrossRef]
84. Ojala, A.E.K.; Kosonen, E.; Weckström, J.; Korkonen, S.; Korhola, A. Seasonal formation of clastic-biogenic varves: The potential for palaeoenvironmental interpretations. *GFF* **2013**, *135*, 237–247. [CrossRef]
85. Saarni, S.; Lensu, A.; Tammelin, M.; Haltia, E.; Saarinen, T. Winter climate signal in boreal clastic-biogenic varves: A comprehensive analysis of three varved records from 1890 to 1990 AD with meteorological and hydrological data from Eastern Finland. *GFF* **2017**, *139*, 314–326. [CrossRef]
86. Boehrer, B.; Schultze, M. Stratification of lakes. *Rev. Geophys.* **2008**, *46*. [CrossRef]
87. Vassiljev, J. The simulated response of lakes to changes in annual and seasonal precipitation: Implication for Holocene lake-level changes in northern Europe. *Clim. Dyn.* **1998**, *14*, 791–801. [CrossRef]
88. Lavapuro, M.; Lipponen, A.; Artimo, A.; Katko, T.S. Groundwater sustainability indicators: Testing with Finnish data. *Boreal Environ. Res.* **2008**, *13*, 381–402.

quaternary

MDPI

Review

Using Annual Resolution Pollen Analysis to Synchronize Varve and Tree-Ring Records

Martin Theuerkauf [1,*], Eike Engelbrecht [1], Nadine Dräger [2], Michael Hupfer [3], Almut Mrotzek [1], Anja Prager [1] and Tobias Scharnweber [1]

1. Department of Peatland Studies and Palaeoecology, Institute of Botany and Landscape Ecology, University of Greifswald, 17489 Greifswald, Germany
2. Section 5.2: Climate Dynamics & Landscape Evolution, GFZ—German Research Centre for Geosciences, Helmholtz Centre Potsdam, 14473 Potsdam, Germany
3. Department of Chemical Analytics and Biogeochemistry, Leibniz-Institute of Freshwater Ecology and Inland Fisheries, 12587 Berlin, Germany

* Correspondence: martin.theuerkauf@uni-greifswald.de; Tel.: +49-3834-4204179

Received: 16 May 2019; Accepted: 3 July 2019; Published: 8 July 2019

Abstract: Fossil wood and varved lake sediments allow proxy analysis with exceptionally high, (sub-)annual resolution. Both archives provide dating through ring and layer counting, yet with different accuracy. In wood, counting errors are small and can be eliminated through cross-dating because tree-rings show regionally synchronous patterns. In varved sediments, counting errors are larger and cross-dating is hampered by missing regional patterns in varve parameters. Here, we test whether annual pollen analysis is suited to synchronize varve records. To that end, annual pollen deposition was estimated in three short cores from two lakes in north-eastern Germany for the period 1980–2017 CE. Analysis has focused on *Fagus sylvatica* and *Picea abies*, which show the strongest annual variations in flowering (mast). For both tree taxa, annual flowering variations recorded by forest and pollen monitoring are well represented in varved lake sediments, hence indeed allow us to synchronize the records. Some pollen mast events were not recognized, which may relate to sampling uncertainties, redeposition or regional variations in flowering. In *Fagus sylvatica*, intense flowering limits wood growth in the same year. Peaks in pollen deposition hence correlate with minima in tree-ring width, which provides a link between varved lake sediments and fossil wood.

Keywords: annually laminated lake sediments; dating; mast; pollen analysis; tree-rings; varves

1. Introduction

Annually laminated (varved) lake sediments, like tree-ring records, are invaluable archives for palaeoecological and palaeoclimatological research with annual or even seasonal resolution. They allow the study of, for example, short-lived effects of volcanic eruptions, pathogen outbreaks, forest fires, landscape clearing, species migration or weather extremes. Accurate dating provided, the study of multiple sites along climate gradients and comparison with other accurately dated, high resolution proxies from, e.g., ice cores or tree-ring studies, has the potential to explore leads and lags of climate and environmental change.

However, such time-sensitive applications rely on accurate dating, ideally to the year, which is commonly achieved in tree-ring but not in varve studies. In tree-ring studies, counting errors are commonly small because ring boundaries are usually well-defined. Errors can occur when under very harsh and unfavourable growing conditions trees have only developed very narrow or occasionally no rings at all, the so-called missing rings [1]. On the other hand, intra-annual density fluctuations may be mistaken for annual ring boundaries, so-called false rings [2]. Such counting errors can be detected/eliminated by careful cross-dating against other, overlapping tree-ring records from the region. Cross-dating is possible because variations in tree-ring width are largely determined by weather conditions during or before the growing season and hence are synchronous across regions, at least for each species [3]. Tree-ring patterns may differ between species because the ecological niches, climatic requirements and physiological strategies of tree species differ. For example, in north-central Europe the main limiting factor for radial growth of many broadleaved species including beech (*Fagus sylvatica*) is summer moisture availability whereas pine (*Pinus sylvestris*) growth is sensitive to late winter/spring temperatures. Cross-dating is also used to match tree-ring sequences from living trees with sequences from fossil wood preserved in, e.g., lakes, peatlands or buildings. With sufficient fossil wood available, long tree-ring chronologies spanning several millennia can be constructed [4–7].

Errors in varve counting are clearly higher than errors in tree-ring counting because the seasonal layers are usually less well defined. Furthermore, varve formation is more easily disturbed either due to changes in seasonal sedimentation or to oxygenation of bottom waters. Seasonal deposition, e.g., diatoms in spring and carbonates in summer, may change by orders of magnitude from year to year due to variable weather conditions [8]. Sedimentation may also vary due to fluctuations in lake mixing. Other than in tree-rings, sediments may be redeposited. As a result, the presence of poorly visible, missing or wrong varves is much higher than similar problems in wood. Hence, errors in varve counting may approach 5% even in well-preserved laminations [9], and higher errors are to be expected with poor varve preservation. Validation with parallel cores from the same site may reduce errors but is rarely applied because of the high work load [9]. Cross-dating between sites is hampered because varve thickness (or other varve parameters), unlike tree-ring width, does not show regionally synchronous variations. While the formation of seasonal layers may well be influenced by weather conditions [8], overall sedimentation is determined by complex interactions between water circulation, lake stratification, biomass productivity, food webs, erosion, etc. Variations in varve thickness are hence site specific. Accurate synchronization of varve records is possible with tephra layers, yet only a limited number of tephra layers, often less than one per millennium, is available even in well studied areas, e.g., Europe [10–12], Japan [13] or New Zealand [14]. Furthermore, recently proof has been provided for ^{10}Be as a novel synchronization tool for varved lake sediments [8,15].

Here, we explore whether annual pollen analysis can improve synchronization and finally dating of varved lake records. The idea is based on the observation that annual pollen deposition of beech in pollen traps, which measure atmospheric pollen deposition, varies by three orders of magnitude [16,17]. The observed variations in pollen deposition relate to annual variations in flowering and seed production (mast), which are an adaption to reduce seed predation and hence to optimize sexual reproduction. Mast years, i.e., years with strong flowering and high seed production, are commonly related to particular weather conditions. Among major European forest trees, such weather cues are spatially consistent only for beech, and to a more limited extent, spruce (*Picea abies*) [18]. For beech, the occurrence of mast years is related to summer temperatures in the two previous years, i.e., intense flowering is triggered by a sequence of a cool–wet summer followed by a warm–dry summer [19–21]. Similarly, for spruce, summer temperatures in the two previous years can explain some variation in seed production, again with a cool summer followed by a warm and dry one triggering cone production in year three [22].

The primary question of the present study is whether the high annual variations in pollen deposition from beech and spruce observed in pollen traps are also recognizable and hence useful in varved lake sediments. To this end we apply annual pollen analysis in three short cores from two lakes in north-eastern Germany. Furthermore, we compare annual pollen deposition with tree-ring records from the same area. Intense flowering and seed production require a substantial amount of resources, so that in mast years fewer resources are available for radial growth, and even less when mast years coincide with summer droughts [23]. In long time series, this effect is visible as an inverse correlation between mast years and ring-width chronologies [24]. We explore whether such a relationship is also recognizable between annual pollen deposition and tree-ring width.

2. Materials and Methods

The present study explores whether annual variations in flowering of beech and spruce are represented and recognizable in varved lake sediments. To that end we study pollen deposition in three cores from surface sediments of two lakes in north-eastern Germany. Our primary study site is Lake Tiefer See, because here modern varve formation has been recently studied in several research projects [8,25,26]. Lake Arendsee was selected for comparison as the nearest site with suitable contemporary varves. The pollen records are compared with forest monitoring data of flowering and seed production, pollen trap data and tree-ring data.

2.1. Study Sites and Coring

Lake Tiefer See, in the centre of Mecklenburg-Western Pomerania, is a deep lake in a glacial meltwater valley (Figures 1 and 2). The lake is part of the Klocksin lake chain. In the north, it receives inflow during wet periods from Lake Flacher See. In the south, a permanent connection exists towards Lake Hofsee. Lake Tiefer See has a surface area of ~0.75 km^2 and a maximum water depth of 62.5 m, which makes it one of the deepest lakes in the lowlands of Northern Germany. The lake is surrounded by a narrow fringe of trees and bushes dominated by *Alnus glutinosa*, *Fraxinus excelsior*, *Quercus petraea*, *Q. robur* and *Corylus avellana*. Beyond that fringe, the lake is surrounded by arable land and open wetlands. Forests in the vicinity of the lake are dominated by *Pinus sylvestris*, while both *Fagus sylvatica* and *Picea abies* are rare within a 10 km distance. The Holocene sediments of Lake Tiefer See show several sections with well-preserved annual laminations (=varves), which however cease during the Medieval period around 1200 CE [26]. Varves are again recognizable in the surface sediments starting at about the year 1924 CE. Dating and monitoring have proven that these varves are indeed annual layers [25,26]. Modern varve formation was probably triggered by the construction of a railway-dam that modified the inflow from Lake Flacher See, and by higher nutrient loads from artificial fertilizers. Higher nutrient loads increase primary production of mainly diatoms and other single-celled algae in the epilimnion of the lake. The decomposition of dead algae leads to increased oxygen consumption and finally the formation of anoxic zones in the deepest part of the lake during summer, which is a precondition for varve formation [8]. The modern varves in Lake Tiefer See are calcite varves with a diatom layer deposited in spring, a calcite layer deposited in summer and an organic layer deposited in autumn [25,26]. Due to higher biomass production, the modern varves are much thicker (~4 mm) than varves deposited before the Medieval period (~1–2 mm).

Lake Arendsee is a karstic lake in northern Saxony-Anhalt (Figures 1 and 2). The main lake basin formed already during the Late Glacial period, and was reshaped in further collapses in 822 CE and 1685 CE [27]. Today, the lake has a surface area of 5.14 km^2, a maximum depth of 48.7 m and a mean depth of 29 m. The lake is mostly surrounded by arable land in the south and west and by pine forests and wetlands in the north and east. *Fagus sylvatica* and *Picea abies* are again rare in the vicinity of the lake. Like in Lake Tiefer See, high nutrient loads and the subsequent appearance of bottom water anoxia has induced the formation of varves since the mid-20th century [27]. The varves are characterized by a calcareous layer, precipitated in June to July, embedded in the algal gyttja [27]. A particular feature of the surface sediments is a massive, ~5 cm thick calcareous layer about 10 cm below the sediment surface during coring in 2017. The layer originates from artificial dispersion of calcareous lake marl extracted from the littoral zone across the lake in the autumn of 1995, which aimed to reduce internal phosphorus release by capping of the sediment [28]. However, lake marl application did not result in any significant decrease of phosphorus loads and the trophic level [29]. We counted 22 calcareous layers above this artificial layer, which indicates that each layer corresponds to one summer between 1996 and 2017, confirming that the varves are indeed annual layers.

Figure 1. Map of the study area, indicating Lake Tiefer See and Lake Arendsee, and position of the pollen traps in Delmenhorst, Vilm, Eldena and Lüssow, as well as position of tree-ring study sites. Green areas indicate forest regions, for which flowering data are recorded.

For the present study we analysed pollen deposition in three short cores, two from Lake Tiefer See and one from Lake Arendsee. Sediment cores of 60 mm diameter were taken with a gravity corer (UWITEC, Mondsee, Austria). The two cores from Lake Tiefer See (TSK15-K1, TSK15-K7) were taken in the deepest lake basin at about 62.5 m water depth in 2015. The core from Arendsee (ARS17A) was taken in the northern part of the main basin at a water depth of about 40 m in 2017.

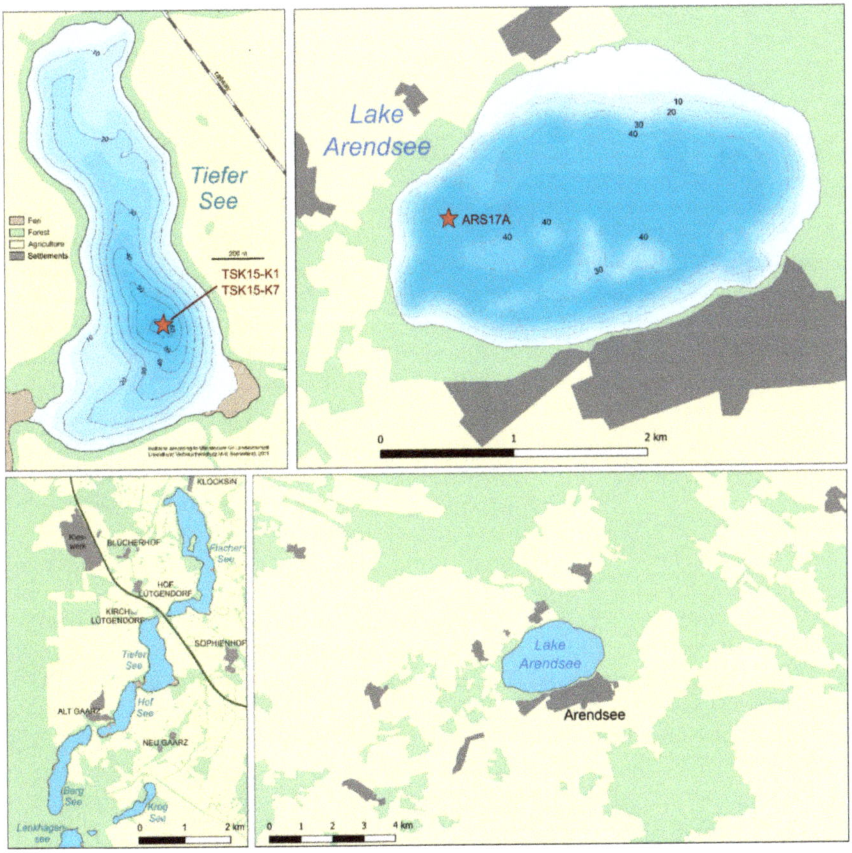

Figure 2. Detailed maps of Lake Tiefer See and Lake Arendsee.

2.2. *Pollen Sampling, Preparation and Analysis*

Sampling annual layers in lake sediments is difficult for common varves with less than 1 mm thickness. The varves in our surface sediments are instead mostly 2–4 mm thick and allowed us a simple sampling strategy.

Cores from Tiefer See remained in the field for some weeks before transport to the lab to allow sediments to consolidate. The cores were opened and kept cool and dry for about one week before sampling to allow further consolidation. Finally, a 1 × 1 cm longitudinal section was extracted from both cores using a u-channel. From this strand, samples were cut off with a sharp blade directly above the calcareous layer of each varve. In the study area, beech flowers in April and May, spruce in May and June. The calcareous layer is deposited from April to July or longer [25]. Hence by cutting just above the calcareous layer we aimed for samples that include the total pollen deposition of Fagus and Picea from the respective year. In all cores, the chronology is based on layer counting during sampling, performed by two analysts (Table 1). For core TSK15-K7, dating is supported by microscopic varve counts on thin sections from the same core [26].

Table 1. Lake surface sediment cores sampled for annual resolution pollen analysis.

Core	Location	Sample Period
TSK15-K1	Lake Tiefer See (53.59° N, 12.53° E)	1981–2015
TSK15-K7	Lake Tiefer See (53.59° N, 12.53° E)	1980–2007
ARS17A	Lake Arendsee (52.8891° N, 11.4598° E)	1983–2017

Pollen sample preparation followed the standard procedure described by Fægri et al. (1989) and includes treatment with 25% HCL, 10% KOH and 7 min acetolysis at 100 °C. One or two tablets with exotic marker grains (*Lycopodium clavatum* spores, Batch Nr. 3862) were added to estimate pollen accumulation. Treatment with hydrofluoric acid (HF) was only applied for samples of core TSK15-K7. Finally, samples were washed with either ethanol or 2-propanol (isopropyl) and transferred to glycerin (TSK15-K7) or silicon oil (TSK15-K1, ARS17A), respectively. Pollen samples were analysed at 400× magnification with a Zeiss-Axiolab microscope. In samples from core TSK15-K7 complete pollen spectra were analysed. Pollen percentages were calculated on the basis of an upland pollen sum. In samples from TSK15-K1 and ARS17A only Fagus (and Picea) pollen as well as exotic marker grains were counted until a minimum of 50 marker grains. The results are presented as annual pollen accumulation rates. To clearly separate between data from plant observations and pollen analysis we set names of pollen types in capital letters [30].

2.3. *Identifying Pollen Mast Years*

To identify mast intensity in the study area since 1980, we compiled information from three monitoring sources: pollen trap records, flowering intensity observations and fructification observations.

Pollen trap data appear to be the best reference for pollen deposition in lakes. For beech, the longest record available from northern Germany is the 32-year record (1982–2014) from Delmenhorst, about 200 km west of our study sites (Figure 1). Here, atmospheric pollen concentration is measured with a Hirst volumetric trap. The record is published as an annual pollen index, which for each year is calculated as the sum of all daily means of Fagus pollen grains per cubic metre of air [17]. In our study area of north-eastern Germany, pollen trapping has been ongoing since 2005 at several locations. Here, we use data from three modified Tauber traps [31] (Table 2). In addition to the original design, a galvanized wire mesh with 1-cm grid size is added below the opening to prevent larger animals from entering the trap. The trap in Lüssow is attached to a tree 80 cm above the ground. The traps Eldena and Vilm are installed on the ground, with the opening 12 cm above the surface. The traps are harvested yearly in late September/early October. Samples were sieved (1-mm mesh size) and prepared following the above protocol. HF was applied for samples with high mineral content. Pollen values for these traps are presented as relative values related to a sum of Pinus and Quercus.

Table 2. Location and details of the pollen trap records used in the present study.

Trap	Location	Site Type	Sampling Period
Delmenhorst	53.05° N 8.63° E	roof top in the city centre, 17.5 m above street level	1982–2014 (1992 missing)
Lüssow	53.8911° N 13.4781° E	Alder (*Alnus glutinosa*) carr (100 ha) in a large river valley mire	2005–2016 (2006 and 2007 missing)
Eldena	54.0784° N 13.4767° E	old-growth forest with *Fagus sylvatica*, *Quercus robur*, *Acer* spp. and *Carpinus betulus*	2014–2018
Vilm	54.3273° N 13.5396° E	old-growth forest with *Fagus sylvatica*, *Quercus robur*, *Acer* spp. and *Carpinus betulus*	2015–2018

Flowering intensity of the major forest trees, including beech and spruce, has been reported for forest regions across Germany since 1992 [32]. Flowering is classified into four categories, from 1 (no or very weak flowering) to 4 (rich flowering). To explore regional variations, here we include data from three forest regions in northern Germany (Figure 1), as well as mean values for northern and southern Germany.

Finally, extensive and long observations exist for fructification, i.e., the seed production of trees. However, fructification is not a fully accurate proxy for flowering and pollen production in the same year because seed development may be interrupted during and after flowering by calamities, late frosts or other weather extremes. Hence, fructification data is primarily used to fill gaps in the other proxies. We used data compiled by federal forest monitoring from the MASTREE database [32].

2.4. Tree-Ring Chronologies

For comparison we use composite tree-ring chronologies of beech from the Müritz area, close to Lake Tiefer See, and from Boizenburg/Schwerin, about 100 km north of Lake Arendsee (Figure 1). The Müritz composite chronology is created from eight sites of mature closed canopy forest stands. Sample replication is 160 trees with a minimum age of 80 years. The Boizenburg/Schwerin chronology comprises 45 trees from two sites with a strong common signal. All chronologies were detrended using flexible spline curves to remove the inherent age trend and possible influences of disturbances or forest management. The resulting index chronologies are thought to best represent the inter-annual variations in tree-growth caused by climate and allocation to reproduction (mast).

2.5. Meteorological Data

The flowering intensity of beech in one year has been attributed to summer temperatures in the two previous years [19–21]. To validate this relationship for the study sites, we compare flowering data with the July temperature offset ($\Delta_{temp}July_{1-2}$), i.e., mean July temperature one year ago minus mean July temperatures two years ago. Mean July temperatures were obtained from Deutscher Wetterdienst (DWD) for the weather station Waren, about 10 km south-east of Lake Tiefer See.

2.6. Statistical Analyses

To compare the pollen records with monitoring data and the tree-ring chronologies we tested for gleichläufigkeit (glk) and correlation using the R environment for statistical computing (R Core Team 2018). Gleichläufigkeit was calculated with the glk function from the 'dplR' package, version 1.6.6 [33–35].

3. Results

3.1. Pollen Mast Years between 1980 and 2018

In this section we compare all monitoring data to identify years with widespread intense flowering and pollen emission. For beech, the pollen traps from Delmenhorst, Lüssow, Eldena and Vilm show sharp, and largely synchronous variations in annual pollen deposition (Figure 3). Since 1982, the Delmenhorst record shows seven years with very high pollen deposition (pollen index > 500: 1983, 1995, 2000, 2004, 2007, 2009, 2011), and five years with intermediate pollen deposition (pollen index between 250 and 500: 1987, 1990, 1998, 2006 and 2014). The pollen traps from north-eastern Germany show very high pollen deposition in 2014 and 2016, so that we overall assume nine years with very high and four years with intermediate pollen deposition. The trap data correspond well to monitoring data of flowering intensity across northern Germany—in all years with intermediate to high pollen deposition since 1995, mean flowering intensity is higher than 2.5 in all sub-regions, indicating intermediate to strong flowering (Figure 3). Flowering intensity is also high in the year 1992, which is missing from the pollen trap record, hence adding a year with presumed high pollen deposition to the record. In only two years we do observe regional variation in the flowering data. In 2002, only region 2 and 3 indicate intense flowering, in 2018 only region 3 indicates intense flowering.

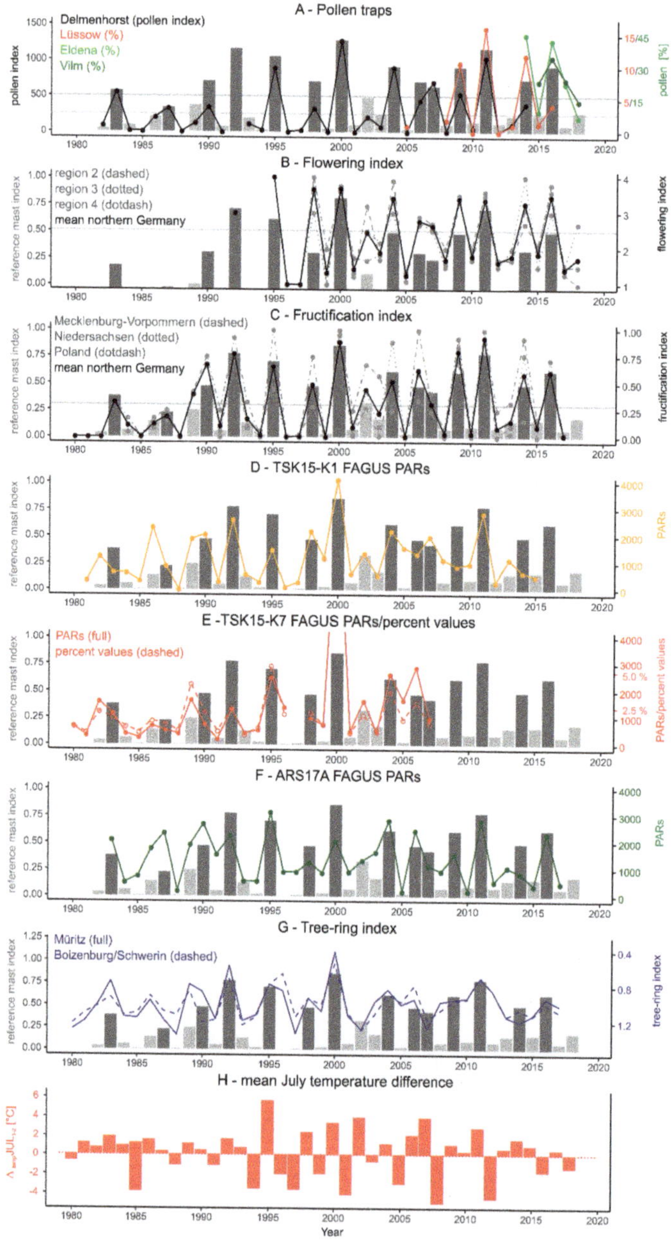

Figure 3. Pollen and forest monitoring data for beech (**A–C**), annual pollen deposition in three sediment cores (**D–F**), ring width index from the Müritz and Boizenburg/Schwerin composite (**G**) and difference in mean July temperature one year ago minus mean the value two years ago (**H**). Grey bars: reference mast index, dark bars indicate supposed pollen mast years. The pollen index (**A**) for each year is the sum of all daily means of Fagus pollen grains per cubic metre of air. Flowering and fructification index (**B,C**) are assigned by forest monitoring. The reference index is calculated as a mean of pollen, flowering and fructification index. Pollen accumulation rates (PARs) (**D–F**) are given in pollen grains cm^{-2} yr^{-1}.

Finally, observed fructification also corresponds well to pollen deposition and flowering: in all years with intermediate to high pollen deposition and flowering (except 1987), the fructification index is higher than 0.3 (Figure 3). Such a high index is also observed in 1989, with pollen index and flowering index being low, and 2002, with pollen index and flowering index just below the threshold for intermediate flowering. Again, the regional records are closely correlated. Only in 1998, 2006 and particularly in 2014, the index is clearly lower in Mecklenburg-Vorpommern than in the other regions.

Overall, we observe ten years with supposedly strong, four years with intermediate and two years with possibly also elevated flowering and hence pollen deposition. Flowering and pollen production has been highly synchronous across northern Germany, and intense flowering in almost all cases resulted in strong fructification. Only in 2014, fructification was low in north-eastern Germany although strong flowering is recorded. For this year, late frost was reported, which may have hampered fructification [36]. For further analysis we use a mean mast index composed of pollen, flowering and fructification index as a reference. For calculation of this mean index, all indices were rescaled to the range 0 to 1.

Comparison with mean July temperatures largely confirms the earlier observation that intense flowering in beech is related to a positive July temperature difference in the two previous years, i.e., when mean July temperatures were low two years ago and high one year ago (Figure 3). However, in 2016 intense flowering is observed although the July temperature difference was negative. This observation may indicate that flowering intensity is influenced by further factors.

For spruce, only short pollen trap records are available. They show high pollen deposition in 2009, 2011, and 2015, and somewhat elevated values in 2013 and 2014. Monitoring of flowering intensity started in 1995 and shows more regional variation for spruce than for beech. In several years, intermediate to intense flowering (index > 2.5) is only recorded in one of the three regions: only in region 4 in 2000 and 2003, only in region 3 in 2008 and only in region 2 in 2013 and 2017. In eight years, all regional records indicate intermediate to strong flowering: 1992, 1998, 2004, 2006, 2009, 2011, 2015 and 2018. For the overlapping period, intense flowering matches high pollen deposition in the pollen traps. However, pollen deposition was low in 2018 although intense flowering was reported. Spring and summer of 2018 were exceptionally dry, which may have limited pollen production per single flower. In another four years, the mean flowering index is above 2.5: 2000, 2003, 2014 and 2016. For fructification, only a mean German record and a record from north-eastern Poland are available. Both largely match the flowering index, with the exception that the mean German record indicates strong fructification in 1993 and 2007, although flowering has been low. The Polish record shows strong fructification also in 1990, which hence may have been a year with intense flowering as well. Widespread intermediate to intense flowering is indicated in seven years, regional flowering in another nine years. As a reference, widespread flowering years are indicated in Figure 4 with a value of 1, regional flowering years with a value of 0.5.

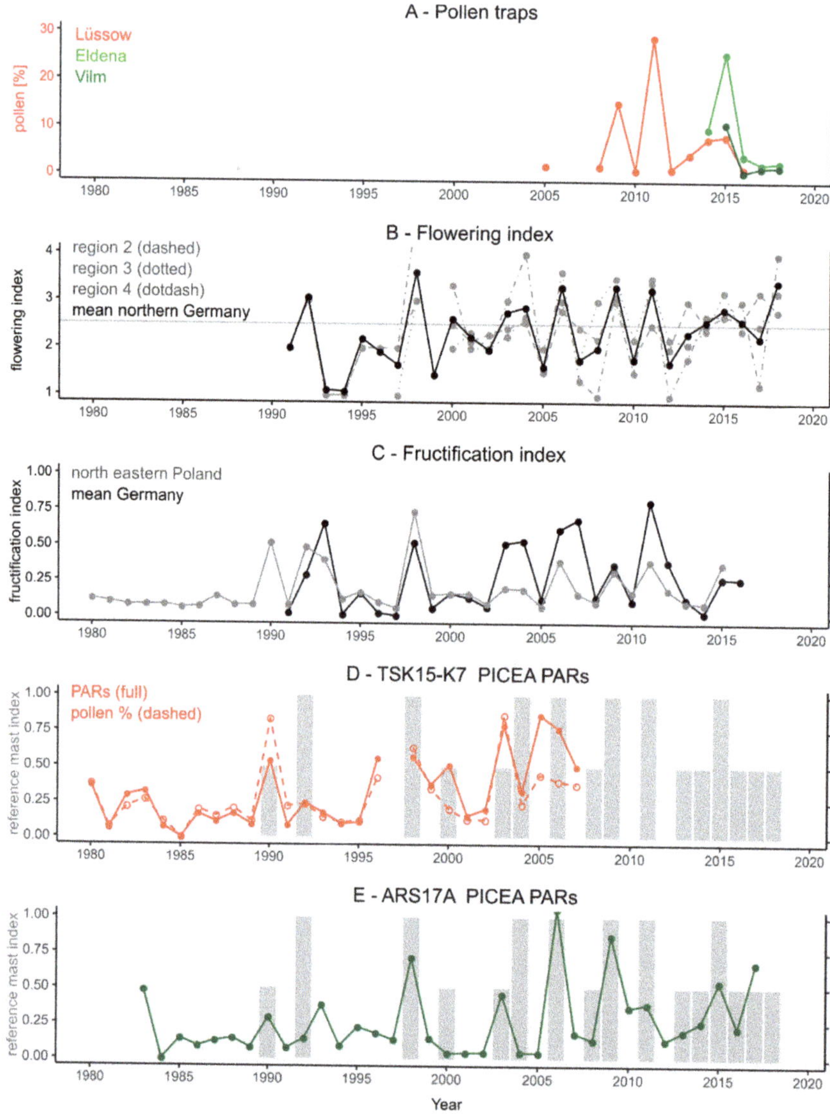

Figure 4. Pollen and forest monitoring data for spruce (**A–C**), annual pollen deposition in three sediment cores (**D–E**). Grey bars: Long bars indicate supposed intense, short bars intermediate flowering years.

3.2. *Pollen Mast Years in the Lake Records*

3.2.1. Fagus

Annual pollen accumulation rates (PARs) of Fagus in the three lake pollen records show high annual variations mostly between ~100 and ~4000 grains cm^{-2} year^{-1} (Figure 3). An exceptionally high value of 9300 grains cm^{-2} year^{-1} is observed in TSK15-K7 for the sample of the year 2000. Also other pollen types in that sample have exceptionally high values, which may indicate a sampling error or an error in PAR estimation. All three lake pollen records closely correlate with the reference mast index, with correlation being strongest for ARS17A (Figure 5, Table 3). To compare the number of pollen mast years detected in the lake records with those observed in the monitoring data, ad-hoc classification was applied: A sample is classified as a pollen mast year if the PAR value in this sample is higher than the PAR value in the sample above or below plus the median over all samples in that record. The pollen mast years identified in this way correspond well to the monitoring data; all pollen mast years identified in the monitoring data, except 2007, were recorded in at least one of the lake pollen records (see Supplementary Materials). However, in each record several pollen mast years have not been identified (false negatives): six in TSK15-K1 (1983, 1987, 2006, 2007, 2009 and 2014), four in TSK15-K7 (1987, 1990, 1998 and 2007) and three in ARS17A (1998, 2014 and 2007). On the other hand, only three pollen mast years have been wrongly detected (false positives), one in TSK15-K1 (1986) and two in TSK15-K7 (1982, 2002).

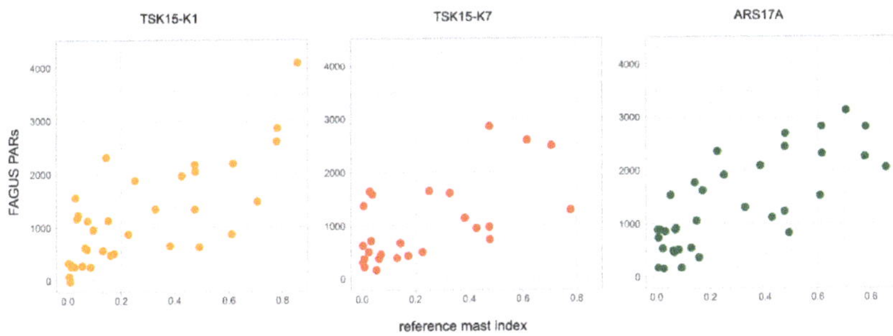

Figure 5. Scatter plots of annual FAGUS PARs over the reference mast index.

Table 3. Correlation coefficients and gleichläufigkeit between annual pollen deposition of FAGUS and PICEA in three lake pollen records and the reference mast index as well as the tree-ring records.

Relationship Tested	TSK15-K1	TSK15-K7	ARS17A
FAGUS PARs ~ reference mast index	$r = 0.73$ $p = 1.124 \times 10^{-6}$ glk = 0.79 (33 a)	$r = 0.63$ $p = 0.00074$ glk = 0.79 (25 a)	$r = 0.77$ $p = 6.133 \times 10^{-8}$ glk = 0.90 (35 a)
PICEA PARs ~ reference mast index	-	$r = 0.36$ $p = 0.1702$ glk = 0.66 (15 a)	$r = 0.56$ $p = 0.002192$ glk = 0.73 (27 a)
Tree-ring index ~ FAGUS PARs	$r = -0.67$ $p = 1.129 \times 10^{-5}$ 1-glk = 0.65 (35 a)	$r = -0.67$ $p = 0.00014$ 1-glk = 0.85 (25 a)	$r = -0.44$ $p = 0.0082$ 1-glk = 0.72 (35 a)

In TSK15-K7, full pollen spectra were counted so that pollen percentages of FAGUS are also available. PARs and pollen percentages show parallel variations, and hence suggest that either data type is suitable to recognize pollen mast years in the pollen record (Figure 3).

3.2.2. PICEA

Annual pollen deposition of PICEA in TSK15-K7 and ARS17A shows similarly high variations as pollen deposition of FAGUS (Figure 4). Furthermore, pollen deposition in Lake Arendsee is ~5 times higher (maximum = 4200 grains cm^{-2} $year^{-1}$) than in Lake Tiefer See (maximum = 700 grains cm^{-2} $year^{-1}$). The higher values in Llake Arendsee probably reflect the smaller distance to extended spruce forests. Spruce is a rare forest tree in Mecklenburg-Vorpommern. Larger stands occur in Lower Saxony, about 75 km west of Lake Arendsee and 200 km south-west of Lake Tiefer See, and in the Harz Mountains, about 150 km south of Lake Arendsee and 250 km south-west of Lake Tiefer See.

Of the seven mast years indicated by the monitoring data, five are recognizable in the pollen record of ARS17A (1998, 2006, 2009, 2011 and 2015) while two years (1992, 2004) do not show elevated PARs.

A similar situation with intense flowering but low pollen deposition in the pollen traps is observed in 2018. The growing season of this year has been exceptionally dry. So, a possible explanation for lower than expected pollen deposition is that pollen development was hampered by the drought. Precipitation has been at average levels in 2004, so that lower than expected pollen deposition in this year may relate to another effect. Already 2003 was a year with elevated flowering and fructification. Due to resource depletion, spruce may not produce abundant pollen in two subsequent years, even if flowers are present. Both effects are only speculative and need verification. On the other hand, pollen deposition of Picea was elevated in 1993 despite low observed flowering in northern Germany. However, high values of the mean German fructification record suggest intense mast in central and southern Germany. Mast is also indicated in north-eastern Poland. The higher than expected pollen deposition in Lake Arendsee may hence relate to long-distance transport of pollen from these areas. Similarly, pollen deposition was elevated in 2017 although intense flowering was only recorded in region 2. Finally, pollen deposition was high in 2010, although low flowering and fructification was reported in all areas. The high PARs for Picea in 2010 may represent redeposition from 2009, which was a strong flowering year. No redeposition is indicated for beech pollen; PARs were high in the mast year 2009 and very low in 2010. However, redeposition may here play a larger role in spruce pollen because this species flowers about four weeks later than beech and the pollen grains float longer. Both effects increase chances that some pollen is deposited at the lake bottom later in the year and hence within the detrital layer that due to the chosen sampling procedure is attributed to the next year.

Only two of the seven mast years indicated in monitoring data are detected in the Lake Tiefer See pollen record TSK15-K7 (1992 and 2006). High deposition is also observed in three years with regional flowering (1990, 2000 and 2003) and three years with only weak flowering (1996, 2005 and 2007). We suggest that the weak link between monitoring data and pollen deposition in Lake Tiefer See is related to the rareness of spruce in the area. Due to the overall low pollen deposition, localized flowering events and random, long distance transport events may strongly influence pollen deposition in that lake.

3.3. *Tree-Ring Records*

The tree-ring indices from the Müritz area and Boizenburg/Schwerin are closely correlated during the study period 1980–2017, showing very similar growth trends in both areas (Figure 3). The index ranges from 0.4 to 1.4, with the lowest values being observed during the years with most intense flowering and highest fructification: 1993, 1995, 2000 and 2010 (Figure 3). Correlation between monitoring data of fructification and the Müritz tree-ring index (1991–2016) is high ($r = -0.67$) and significant. Similarly we found close negative relationships between annual pollen deposition in the three pollen records and the tree-ring indices (Figure 6, Table 3). The correlation is somewhat closer between the two Tiefer See pollen records and the Müritz tree-ring index ($r = -0.67$ for both records) than between the Arendsee pollen record and the Boizenburg/Schwerin tree-ring index ($r = -0.44$). These results indicate that tree growth is depressed during years of high pollen production. Such a negative relationship between reproduction and growth is well known for beech stands across Europe [19,21,37]. It has been interpreted as a trade-off in resource allocation between reproduction on the one side and radial tree growth on the other [19,37,38]. In addition, beech as a flowering masting species [39] shows a high coherence between flowering intensity and final seed production, i.e., the number of seeds produced is largely controlled by the flowering effort. In fruit-maturation masting species like oak (*Quercus* spp.), fruit production is instead more strongly controlled by the variable ripening of a more constant flower crop [39]. For flowering masting species, a high coherence between pollen production and final seed crop can thus be expected, which is supported by our results. Only in years with unfavourable spring/early summer conditions (strong drought) or in years with late frost events massive flowering will not result in high seed crops [40,41].

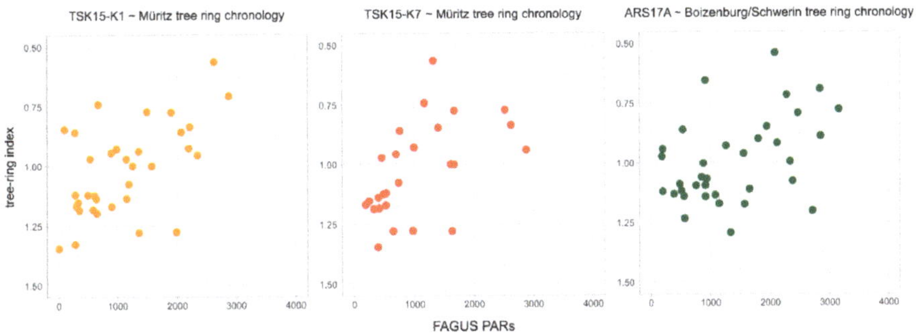

Figure 6. Scatter plots of tree-ring width in the Müritz and Boizenburg/Schwerin composite chronologies over annual Fagus PARs.

4. Discussion

4.1. Annual Pollen Data

The close correlation between the mean masting index and annual pollen deposition in varved lake sediment cores shows that annual variations in pollen production of beech and spruce are indeed recognizable in annually laminated sediments, although not in all cases. For beech all the years with strong flowering have been well identified, yet some of the years with intermediate flowering are missing in one or several lake records, e.g., 2007 and 2014. Only in 2006 and 2007 is intermediate flowering observed in two consecutive years. In 2007, pollen production may have been low despite a high number of observed flowers, e.g., due to resource depletion. For spruce, a similar situation is observed in 2003 and 2004. Alternatively, pollen production and later seed production may have been disturbed by the exceptionally warm and dry conditions during the summer of 2003. In 2018, intense flowering in spruce was recorded yet pollen deposition in the pollen traps remained low. In 2014, fructification in beech was regionally low although flowering was widespread and high. In this case, however, pollen deposition in the pollen traps was also high. Hence, seed production may have been disturbed after flowering, at least in particular areas.

Lower than expected pollen deposition in the lakes, as in 2014, may also be related to sampling uncertainties. The varves in the studied sections were mostly 2–4 mm thick, which allowed us a simple slicing approach. However, particularly in the Lake Tiefer See cores, not all varves are clearly recognizable by the naked eye, and seasonal layers tend to be undulating. Moreover, the surface sediments in both lakes are still soft. Hence, despite the rather large varve thickness, separation of single years may not have been fully successful in all cases. Finally, the pollen signal in lake bottom sediments may be disturbed by redeposition of pollen. In the study area, the main pollen season of beech and spruce occurs between April and June. We have separated samples directly above the distinctive calcareous layer, which is observed to be deposited from April until at least July, but often longer [25]. Hence, the pollen deposition of beech and spruce from one season is expected to be present in the sample of that year. However, pollen initially deposited at the lake margins or in its surroundings may be later redeposited and hence end up in sediments attributed to the following year. In our records, pollen deposition of both trees mostly declines sharply after a year with peak pollen deposition, suggesting that redeposition has only minor effects. Ref. [42] correspondingly found little indication of redeposition in lakes with a large, deep basin and a small littoral zone, like our study sites. Only in some cases, PARs remain high in the year after a mast year, e.g., for Fagus in 1991 (ARS17A) and 2005 (both cores from Tiefer See) and for Picea in 2010 (ARS17A). Redeposition thus may be relevant in some years.

4.2. Outlook

4.2.1. Application of Annual Pollen Records

Flowering and pollen production in beech and spruce vary sharply from year to year. These variations are well represented in pollen traps, and also preserved and recognizable in annually laminated lake sediments. We suggest that such annual pollen counts are potentially useful in three ways.

1. Synchronization and Dating of Varved Sediments

As the main application, we consider annual pollen counts as useful to synchronize varved sediment sections from sites across a region. Annually laminated lake sediments are valuable climate and environmental archives, because they provide up to sub-annual resolution. Exploiting the full potential of varved records requires dating to the exact year, which is potentially provided through layer counting. In reality, however, invisible, blurred, or duplicate layers cause dating errors [9]. Furthermore, many varve records include poorly or non-varved sections, hence creating floating varve sections. So far only tephra layers allow exact synchronization of varve sections for, e.g., transect studies, yet only a few tephra layers are recorded in central Europe [10]. Annual pollen counts instead may provide continuous synchronization. If a sufficient number of long records are available, continuous pollen mast chronologies may be established. Like tree-ring chronologies, they will allow the dating of further (floating) varve sections.

The primary question, however, is whether the pollen mast signal is indeed sufficient to synchronize records. In northern Germany, pollen mast years have occurred regularly since 1990, often every second year. Such a regular pattern provides limited links for synchronization. Long mast records instead show more variation, i.e., clusters of regular flowering and long pauses well suited for synchronization [24]. Robust synchronization requires records with a minimum length. In tree-ring science, typically overlaps of 50 years or more are considered to give reliable cross-dating results. Similar values may apply for mast records. Longer overlap may be necessary in periods with very regular masting. On the other hand, further marker horizons, such as tephra layers, will reduce the necessary overlap. Also analysis of pollen mast in several taxa may reduce the necessary overlap. In addition to *Fagus sylvatica* and *Picea abies*, *Carpinus betulus* and *Abies alba* also show higher flowering variability (Table 4). Whether they provide a meaningful pattern in annually laminated pollen records still needs to be tested. *Quercus robur* and *Q. petraea*—although well known for pronounced seed mast—show low variation in flowering intensity and probably provide no robust pattern in annual pollen deposition. As mentioned before, variable fruit production in *Quercus* is more strongly controlled by the variable fruit ripening rather than by variable flowering.

Table 4. Standard deviation (SD) in the mean flowering index of major forest trees in Germany during the period 1992–2018.

Species	SD
Abies alba	0.52
Alnus glutinosa	0.39
Betula pendula	0.29
Carpinus betulus	0.62
Fagus sylvatica	0.82
Picea abies	0.81
Pinus sylvestris	0.41
Qercus petraea	0.41
Qercus robur	0.39
Tilia cordata	0.37

Synchronization itself may be applied in different ways. Records may be either synchronized using a binary approach, in which each annual sample is classified as mast or non-mast, e.g., in a higher-than-neighbour approach as in the present study. Alternatively, records may be synchronized using their gleichläufigkeit, correlation or the T-values approach. The latter method is used for cross-dating tree-rings. Calculation is based on correlation coefficients between two records, weighted by the length of the overlapping period. Thus, the same correlation values indicate a more reliable match when found in longer records. Although there is no commonly agreed threshold, T-values above 4 are often considered as reliable for cross-dating. For a 50 year overlap a T-value of 4 would correspond to a correlation coefficient of 0.5. The T-value approach will be particularly useful when annual variation in pollen deposition is overlaid by longer term changes in pollen deposition due species expansion or decline.

Synchronization with annual pollen data has spatial limits defined by the area over which trees flower simultaneously. Forest and pollen monitoring show that flowering of beech is indeed synchronous across northern Germany and probably also northern/north-western Poland. Over longer distances, i.e., between northern and southern Germany/Switzerland, several mast years are synchronous (2009, 2011 and 2016), while others were limited to the north (2000, 2004, 2014) or south (2003, 2006, 2018). Hence, both areas would probably need separate pollen mast chronologies. Also Ref. [16], on a N-S transect from Poland to Bulgaria, shows some widely synchronous mast years but also regional differences. Further monitoring data, e.g., those compiled for pollen forecast, will allow defining areas with synchronous variations in flowering and pollen production/deposition.

Application of flowering chronologies will also have temporal limits. In central Europe, beech and spruce show the most prominent variations in annual flowering among all major tree taxa and hence are most suited for the approach. However, in the northern part, beech became an abundant tree only ~2500 years ago while spruce is not native to the area and has only been planted over the past 150 years. To the south both taxa have been present for much longer, so that probably longer mast chronologies covering 6000 years or more may be established. Whether strong mast years in the south are also reflected in pollen records from northern Germany due to long distance transport of pollen still needs to be tested. Such long-distance transport is indicated by continuous yet rare presence of Fagus and Picea in pollen records from northern Germany since about 6000–8000 cal. BP.

Finally, the observed close link between annual pollen accumulation of beech and tree-ring width suggests that varve sequences may be dated through direct linking with tree-ring records. The present sequences are too short to fully validate this option. We therefore aim to extend analysis to past periods with overlapping varve and tree-ring records. For northern Germany, the tree-ring record for beech goes back about 1040 years until 980 CE [43]. In Lake Tiefer See, well varved sections exist before 1200 CE, hence the overlap is about 200 years [26].

2. Tree-Ring Studies

Our results confirm that radial growth in beech is strongly influenced by allocation of carbohydrates to reproduction, i.e., flowering and fructification, which itself is triggered by weather parameters [38,44]. This interplay of climate, reproduction and tree growth implies that climate controls on tree growth are partly indirect and delayed. At least in beech, mast can explain the often-observed influence of weather conditions in the previous year on current year's growth [37]. Disentangling the climatic from the reproduction effect in the analysis of tree-ring width would enhance the quality of tree-ring-based climate reconstructions. The relationship between annual PARs in varves and tree-ring chronologies is therefore a mutual one: First, long tree-ring chronologies can provide a link for long annual PAR records before the era of forest monitoring data, hence help to date varve records. Second, varve-derived records of strong flowering events can help to "clean" tree-ring records from the mast effect—by identifying narrow rings related to mast events—in order to strengthen the inherent climate signal.

3. Climate Reconstruction

In several European tree species, most prominently beech and spruce, masting is highly synchronous and has shown to be triggered by large scale weather patterns, e.g., such as that represented by indices of the north Atlantic circulation (NAO) [45,46]. Reconstructions of mast years may hence be an additional proxy for those indices in the past. For beech, the closest correlation exists between mast and a high positive offset in July temperatures in the two years before mast [18,19,47]. A high mast frequency hence points at high inter-annual variations in summer temperatures while absence of mast instead indicates low variation.

4.2.2. Practical Issues

From previous experience the main practical challenge of annual pollen analysis is sediment sampling. Here, we sampled rather thick varves but even then, sampling was not easy because of several not well-defined varves and soft sediments. In older, more consolidated sediments, varves are commonly as thin as 0.5 mm or less. In this case, different sampling strategies using magnification will be required. We sliced actual samples from a longitudinal core section of about 1.5 × 1.5 cm across, but also smaller sections (e.g., 0.5 × 0.5 cm) would contain a sufficient number of pollen grains while at the same time providing higher sampling accuracy. Using such narrow core sections would also simplify parallel sampling for validation.

A second limitation derives from the large number of pollen samples. Full spectrum pollen analysis is time consuming, and certainly limits analysis of many hundreds or thousands of annual samples. Here, we applied simplified analysis, i.e., only Fagus pollen, Picea pollen and the exotic marker were counted. The exotic marker counts were as high as ~500 for samples from Tiefer See but as low as ~50 in samples from Arendsee, which appears to be sufficient for reliable PAR values because of the high variations. Thus, analysis time was reduced to about 20 min per sample, i.e., analysis of 10–15 samples per day is feasible. These efforts will often remain too high for complete analysis of long varve sections, but may be worth doing when the focus is on shorter time periods.

Supplementary Materials: The following is available online at http://www.mdpi.com/2571-550X/2/3/23/s1, Figure S1: Identification of mast years in monitoring data and lake sediments.

Author Contributions: Conceptualization, M.T. and E.E.; Formal analysis, M.T. and E.E.; Investigation, M.T., E.E., N.D., A.M., A.P. and T.S.; Methodology, M.T., E.E. and T.S.; Project administration, M.T.; Resources, M.H.; Visualization, M.T.; Writing–original draft, E.E.; Writing–review & editing, M.T., N.D., M.H., A.M., A.P. and T.S.

Funding: This study is a contribution to the Virtual Institute of Integrated Climate and Landscape Evolution Analysis—ICLEA—of the Helmholtz Association (VH-VI-415) and to BaltRap: The Baltic Sea and its southern Lowlands: proxy—environment interactions in times of Rapid change (SAW-2017-IOW2). The work of AM was funded by the European Social Fund (ESF) and the Ministry of Education, Science and Culture of Mecklenburg-Western Pomerania (project WETSCAPES, ESF/14-BM-A55-0030/16).

Acknowledgments: We are grateful for very helpful comments of three anonymous reviewers.

Conflicts of Interest: The authors declare no conflict of interest.

References

1. Wilmking, M.; Hallinger, M.; Van Bogaert, R.; Kyncl, T.; Babst, F.; Hahne, W.; Juday, G.; De Luis, M.; Novak, K.; Völlm, C. Continuously missing outer rings in woody plants at their distributional margins. *Dendrochronologia* **2012**, *30*, 213–222. [CrossRef]
2. De Micco, V.; Campelo, F.; De Luis, M.; Bräuning, A.; Grabner, M.; Battipaglia, G.; Cherubini, P. Intra-annual density fluctuations in tree rings: How, when, where, and why? *IAWA J.* **2016**, *37*, 232–259. [CrossRef]
3. Stokes, M.A.; Smiley, T.L. *An Introduction to Tree-ring Dating*; University of Arizona Press: Tucson, AZ, USA, 1996; ISBN 978-0-8165-1680-3.

4. Friedrich, M.; Remmele, S.; Kromer, B.; Hofmann, J.; Spurk, M.; Kaiser, K.F.; Orcel, C.; Küppers, M. The 12,460-Year Hohenheim Oak and Pine Tree-Ring Chronology from Central Europe—A Unique Annual Record for Radiocarbon Calibration and Paleoenvironment Reconstructions. *Radiocarbon* **2004**, *46*, 1111–1122. [CrossRef]
5. Grudd, H.; Briffa, K.R.; Karlén, W.; Bartholin, T.S.; Jones, P.D.; Kromer, B. A 7400-year tree-ring chronology in northern Swedish Lapland: Natural climatic variability expressed on annual to millennial timescales. *Holocene* **2002**, *12*, 657–665. [CrossRef]
6. Leuschner, H.H.; Sass-Klaassen, U.; Jansma, E.; Baillie, M.G.; Spurk, M. Subfossil European bog oaks: Population dynamics and long-term growth depressions as indicators of changes in the Holocene hydro-regime and climate. *Holocene* **2002**, *12*, 695–706. [CrossRef]
7. Pilcher, J.R.; Baillie, M.G.L.; Schmidt, B.; Becker, B. A 7272-year tree-ring chronology for western Europe. *Nature* **1984**, *312*, 150–152. [CrossRef]
8. Kienel, U.; Dulski, P.; Ott, F.; Lorenz, S.; Brauer, A. Recently induced anoxia leading to the preservation of seasonal laminae in two NE-German lakes. *J. Paleolimnol.* **2013**, *50*, 535–544. [CrossRef]
9. Żarczyński, M.; Tylmann, W.; Goslar, T. Multiple varve chronologies for the last 2000 years from the sediments of Lake Żabińskie (northeastern Poland)—Comparison of strategies for varve counting and uncertainty estimations. *Quat. Geochronol.* **2018**, *47*, 107–119. [CrossRef]
10. Wulf, S.; Dräger, N.; Ott, F.; Serb, J.; Appelt, O.; Guðmundsdóttir, E.; van den Bogaard, C.; Słowiński, M.; Błaszkiewicz, M.; Brauer, A. Holocene tephrostratigraphy of varved sediment records from Lakes Tiefer See (NE Germany) and Czechowskie (N Poland). *Quat. Sci. Rev.* **2016**, *132*, 1–14. [CrossRef]
11. Jones, G.; Lane, C.S.; Brauer, A.; Davies, S.M.; de Bruijn, R.; Engels, S.; Haliuc, A.; Hoek, W.Z.; Merkt, J.; Sachse, D.; et al. The Lateglacial to early Holocene tephrochronological record from Lake Hämelsee, Germany: A key site within the European tephra framework. *Boreas* **2018**, *28*, 28–40. [CrossRef]
12. Wastegård, S.; Gudmundsdóttir, E.R.; Lind, E.M.; Timms, R.G.O.; Björck, S.; Hannon, G.E.; Olsen, J.; Rundgren, M. Towards a Holocene tephrochronology for the Faroe Islands, North Atlantic. *Quat. Sci. Rev.* **2018**, *195*, 195–214. [CrossRef]
13. Smith, V.C.; Staff, R.A.; Blockley, S.P.; Ramsey, C.B.; Nakagawa, T.; Mark, D.F.; Takemura, K.; Danhara, T. Identification and correlation of visible tephras in the Lake Suigetsu SG06 sedimentary archive, Japan: Chronostratigraphic markers for synchronising of east Asian/west Pacific palaeoclimatic records across the last 150 ka. *Quat. Sci. Rev.* **2013**, *67*, 121–137. [CrossRef]
14. Zawalna-Geer, A.; Lindsay, J.M.; Davies, S.; Augustinus, P.; Davies, S. Extracting a primary Holocene crytoptephra record from Pupuke maar sediments, Auckland, New Zealand. *J. Quat. Sci.* **2016**, *31*, 442–457. [CrossRef]
15. Czymzik, M.; Muscheler, R.; Adolphi, F.; Mekhaldi, F.; Dräger, N.; Ott, F.; Słowiński, M.; Błaszkiewicz, M.; Aldahan, A.; Possnert, G.; et al. Synchronizing 10Be in two varved lake sediment records to IntCal13 14C during three grand solar minima. *Clim. Past* **2018**, *14*, 687–696. [CrossRef]
16. Pidek, I.A.; Svitavská-Svobodová, H.; Van Der Knaap, W.O.; Noryśkiewicz, A.M.; Filbrandt-Czaja, A.; Noryśkiewicz, B.; Latałowa, M.; Zimny, M.; Święta-Musznicka, J.; Bozilova, E.; et al. Variation in annual pollen accumulation rates of Fagus along a N–S transect in Europe based on pollen traps. *Veg. Hist. Archaeobotany* **2010**, *19*, 259–270. [CrossRef]
17. Simoleit, A.; Wachter, R.; Gauger, U.; Werchan, M.; Werchan, B.; Zuberbier, T.; Bergmann, K.-C. Pollen season of European beech (*Fagus sylvatica* L.) and temperature trends at two German monitoring sites over a more than 30-year period. *Aerobiologia* **2016**, *32*, 489–497. [CrossRef]
18. Nussbaumer, A.; Waldner, P.; Apuhtin, V.; Aytar, F.; Benham, S.; Bussotti, F.; Eichhorn, J.; Eickenscheidt, N.; Fabianek, P.; Falkenried, L.; et al. Impact of weather cues and resource dynamics on mast occurrence in the main forest tree species in Europe. *For. Ecol. Manag.* **2018**, *429*, 336–350. [CrossRef]
19. Drobyshev, I.; Övergaard, R.; Saygin, I.; Niklasson, M.; Hickler, T.; Karlsson, M.; Sykes, M.T. Masting behaviour and dendrochronology of European beech (*Fagus sylvatica* L.) in southern Sweden. *For. Ecol. Manag.* **2010**, *259*, 2160–2171. [CrossRef]
20. Hacket-Pain, A.J.; Friend, A.D.; Lageard, J.G.; Thomas, P.A. The influence of masting phenomenon on growth-climate relationships in trees: Explaining the influence of previous summers' climate on ring width. *Tree Physiol.* **2015**, *35*, 319–330. [CrossRef] [PubMed]

21. Vacchiano, G.; Hacket-Pain, A.; Turco, M.; Motta, R.; Maringer, J.; Conedera, M.; Drobyshev, I.; Ascoli, D.; Hacket-Pain, A. Spatial patterns and broad-scale weather cues of beech mast seeding in Europe. *New Phytol.* **2017**, *215*, 595–608. [CrossRef] [PubMed]
22. Piovesan, G.; Adams, J.M.; Bernabei, M.; Selås, V. Climatic factors controlling reproduction and growth of Norway spruce in southern Norway. *Can. J. For. Res.* **2002**, *32*, 217–225.
23. Hacket-Pain, A.J.; Lageard, J.G.A.; Thomas, P.A. Drought and reproductive effort interact to control growth of a temperate broadleaved tree species (*Fagus sylvatica*). *Tree Physiol.* **2017**, *37*, 744–754. [CrossRef] [PubMed]
24. Drobyshev, I.; Niklasson, M.; Mazerolle, M.J.; Bergeron, Y. Reconstruction of a 253-year long mast record of European beech reveals its association with large scale temperature variability and no long-term trend in mast frequencies. *Agric. For. Meteorol.* **2014**, *192*, 9–17. [CrossRef]
25. Kienel, U.; Kirillin, G.; Brademann, B.; Plessen, B.; Lampe, R.; Brauer, A. Effects of spring warming and mixing duration on diatom deposition in deep Tiefer See, NE Germany. *J. Paleolimnol.* **2017**, *57*, 37–49. [CrossRef]
26. Dräger, N.; Theuerkauf, M.; Szeroczyńska, K.; Wulf, S.; Tjallingii, R.; Plessen, B.; Kienel, U.; Brauer, A.; Szeroczynska, K.; Wulf, S.; et al. Varve microfacies and varve preservation record of climate change and human impact for the last 6000 years at Lake Tiefer See (NE Germany). *Holocene* **2017**, *27*, 1–15. [CrossRef]
27. Scharf, B.W. Eutrophication history of Lake Arendsee (Germany). *Palaeogeogr. Palaeoclim. Palaeoecol.* **1998**, *140*, 85–96. [CrossRef]
28. Stüben, D.; Walpersdorf, E.; Voss, K.; Ronicke, H.; Schimmele, M.; Baborowski, M.; Luther, G.; Elsner, W. Application of lake marl at Lake Arendsee, NE Germany: First results of a geochemical monitoring during the restoration experiment. *Sci. Total. Environ.* **1998**, *218*, 33–44. [CrossRef]
29. Hupfer, M.; Pöthig, R.; Brüggemann, R.; Geller, W. Mechanical resuspension of autochthonous calcite (Seekreide) failed to control internal phosphorus cycle in a eutrophic lake. *Water Res.* **2000**, *34*, 859–867. [CrossRef]
30. Joosten, H.; de Klerk, P. What's in a name? Some thoughts on pollen classification, identification, and nomenclature in Quaternary palynology. *Rev. Palaeobot. Palynol.* **2002**, *122*, 29–45. [CrossRef]
31. Tauber, H. A static non-overload pollen collector. *New Phytol.* **1974**, *73*, 359–369. [CrossRef]
32. Ascoli, D.; Maringer, J.; Hacket-Pain, A.; Conedera, M.; Drobyshev, I.; Motta, R.; Cirolli, M.; Kantorowicz, W.; Zang, C.; Schueler, S.; et al. Two centuries of masting data for European beech and Norway spruce across the European continent. *Ecology* **2017**, *98*, 1473. [CrossRef] [PubMed]
33. Bunn, A.G. A dendrochronology program library in R (dplR). *Dendrochronologia* **2008**, *26*, 115–124. [CrossRef]
34. Bunn, A.G. Statistical and visual crossdating in R using the dplR library. *Dendrochronologia* **2010**, *28*, 251–258. [CrossRef]
35. Bunn, A.G.; Korpela, M.; Biondi, F.; Campelo, F.; Qeadan, F.; Mérian, P.; Zang, C. dplR: Dendrochronology Program Library in R. Available online: https://cran.r-project.org/web/packages/dplR/index.html (accessed on 8 July 2019).
36. Schneck, D. Das Blühen der Waldbäume. *AFZ-DerWald* **2014**, *69*, 27–29.
37. Hacket-Pain, A.J.; Ascoli, D.; Vacchiano, G.; Biondi, F.; Cavin, L.; Conedera, M.; Drobyshev, I.; Liñán, I.D.; Friend, A.D.; Grabner, M.; et al. Climatically controlled reproduction drives interannual growth variability in a temperate tree species. *Ecol. Lett.* **2018**, *21*, 1833–1844. [CrossRef] [PubMed]
38. Mund, M.; Kutsch, W.L.; Wirth, C.; Kahl, T.; Knohl, A.; Skomarkova, M.V.; Schulze, E.D. The influence of climate and fructification on the inter-annual variability of stem growth and net primary productivity in an old-growth, mixed beech forest. *Tree Physiol.* **2010**, *30*, 689–704. [CrossRef] [PubMed]
39. Pearse, I.S.; Koenig, W.D.; Kelly, D. Mechanisms of mast seeding: Resources, weather, cues, and selection. *New Phytol.* **2016**, *212*, 546–562. [CrossRef]
40. Bogdziewicz, M.; Szymkowiak, J.; Fernández-Martínez, M.; Peñuelas, J.; Espelta, J.M. The effects of local climate on the correlation between weather and seed production differ in two species with contrasting masting habit. *Agric. For. Meteorol.* **2019**, *268*, 109–115. [CrossRef]
41. Kasprzyk, I.; Ortyl, B.; Dulska-Jeż, A. Relationships among weather parameters, airborne pollen and seed crops of *Fagus* and *Quercus* in Poland. *Agric. For. Meteorol.* **2014**, *197*, 111–122. [CrossRef]
42. Giesecke, T.; Fontana, S.L. Revisiting pollen accumulation rates from Swedish lake sediments. *Holocene* **2008**, *18*, 293–305. [CrossRef]

43. Scharnweber, T.; Heußner, K.-U.; Smiljanic, M.; Heinrich, I.; Van Der Maaten-Theunissen, M.; Van Der Maaten, E.; Struwe, T.; Buras, A.; Wilmking, M. Removing the no-analogue bias in modern accelerated tree growth leads to stronger medieval drought. *Sci. Rep.* **2019**, *9*, 2509. [CrossRef] [PubMed]
44. Skomarkova, M.V.; Vaganov, E.A.; Mund, M.; Knohl, A.; Linke, P.; Boerner, A.; Schulze, E.-D. Inter-annual and seasonal variability of radial growth, wood density and carbon isotope ratios in tree rings of beech (*Fagus sylvatica*) growing in Germany and Italy. *Trees* **2006**, *20*, 571–586. [CrossRef]
45. Fernández-Martínez, M.; Vicca, S.; Janssens, I.A.; Espelta, J.M.; Peñuelas, J. The North Atlantic Oscillation synchronises fruit production in western European forests. *Ecography* **2017**, *40*, 864–874. [CrossRef]
46. Ascoli, D.; Vacchiano, G.; Turco, M.; Conedera, M.; Drobyshev, I.; Maringer, J.; Motta, R.; Hacket-Pain, A. Inter-annual and decadal changes in teleconnections drive continental-scale synchronization of tree reproduction. *Nat. Commun.* **2017**, *8*, 2205. [CrossRef]
47. Piovesan, G.; Adams, J.M. Masting behaviour in beech: Linking reproduction and climatic variation. *Can. J. Bot.* **2001**, *79*, 1039–1047.

quaternary

MDPI

Article

Dropstones in Lacustrine Sediments as a Record of Snow Avalanches—A Validation of the Proxy by Combining Satellite Imagery and Varve Chronology at Kenai Lake (South-Central Alaska)

Sien Thys [1,*], Maarten Van Daele [1,*], Nore Praet [1], Britta J.L. Jensen [2], Thomas Van Dyck [1], Peter J. Haeussler [3], Elke Vandekerkhove [1], Veerle Cnudde [4] and Marc De Batist [1]

1 Renard Centre of Marine Geology (RCMG), Department of Geology, Ghent University, Krijgslaan, 281 S8 Ghent, Belgium; Nore.Praet@UGent.be (N.P.); thomas-vandyck@hotmail.com (T.V.D.); Elke.Vandekerkhove@UGent.be (E.V.); Marc.DeBatist@UGent.be (M.D.B.)

2 Department of Earth and Atmospheric Science, University of Alberta, Edmonton, AB T6G 2E3, Canada; bjjensen@ualberta.ca

3 U.S. Geological Survey, Anchorage, 4210 University Drive, Anchorage, AK 99508, USA; pheuslr@usgs.gov

4 Centre of X-ray Tomography-PProGRess, Department of Geology, Ghent University, Krijgslaan, 281 S8 Ghent, Belgium; Veerle.Cnudde@UGent.be

* Correspondence: sienthys@gmail.com (S.T.); Maarten.VanDaele@UGent.be (M.V.D.); Tel.: +32-(0)9264-45-73 (M.V.D.)

Received: 12 December 2018; Accepted: 24 February 2019; Published: 1 March 2019

check for updates

Abstract: Snow avalanches cause many fatalities every year and damage local economies worldwide. The present-day climate change affects the snowpack and, thus, the properties and frequency of snow avalanches. Reconstructing snow avalanche records can help us understand past variations in avalanche frequency and their relationship to climate change. Previous avalanche records have primarily been reconstructed using dendrochronology. Here, we investigate the potential of lake sediments to record snow avalanches by studying 27 < 30-cm-long sediment cores from Kenai Lake, south-central Alaska. We use X-ray computed tomography (CT) to image post-1964 varves and to identify dropstones. We use two newly identified cryptotephras to update the existing varve chronology. Satellite imagery is used to understand the redistribution of sediments by ice floes over the lake, which helps to explain why some avalanches are not recorded. Finally, we compare the dropstone record with climate data to show that snow avalanche activity is related to high amounts of snowfall in periods of relatively warm or variable temperature conditions. We show, for the first time, a direct link between historical snow avalanches and dropstones preserved in lake sediments. Although the lacustrine varve record does not allow for the development of a complete annual reconstruction of the snow avalanche history in the Kenai Lake valley, our results suggest that it can be used for long-term decadal reconstructions of the snow-avalanche history, ideally in combination with similar records from lakes elsewhere in the region.

Keywords: dropstones; snow avalanche; X-ray CT; varve

1. Introduction

Over the period 2006 to 2016, an average of 27 people died each year in the United States of America due to snow avalanches [1]. In the European Alps, the number of fatalities is even higher, with a yearly average of 100 casualties over the past 40 years [2]. Furthermore, snow avalanches also have socio-economic consequences, such as road blockages, and power plant and power line damage [3]. Ongoing climate change affects the air temperature in mountainous regions and influences snow conditions, which are key factors in controlling magnitude, frequency, and seasonality of snow avalanches [4]. Snow-avalanche records provide insights into possible triggering mechanisms and the periodicity of events. Triggering mechanisms can be both human induced and climate dependent, and understanding these mechanisms helps to assess the hazard. The more accurate and complete the snow avalanche record is, the better snow-avalanche hazards can be estimated and provide important information for infrastructure development [5].

Most snow-avalanche frequency studies have been based on historic avalanche records and avalanche records from tree rings [6–8]. Trees are often present along snow-avalanche tracks and get impacted (e.g., damaged, inclined, loosening neighboring trees) by the avalanches. The trees react to the changed circumstances, causing variations in the tree rings that can be used to develop an avalanche chronology [8]. Trees along a certain avalanche track will, however, only record information of avalanches passing this particular track and not from avalanches on other tracks. To create the most complete avalanche chronology for a region, data from every avalanche track should, thus, be included.

Lakes, on the other hand, have the potential to catch avalanche deposits from several tracks in the mountains surrounding them. In 1975, Luckman [9] postulated that dirty snow avalanches that run out on lake ice may distribute dropstones over the lake during the melt season. Since then, several studies have shown records of such dropstones, especially in Norwegian lakes [9–11]. Sieving, loss-on-ignition (LOI; for organic matter content) and magnetic-susceptibility (MS) measurements were applied to distinguish the coarser, angular avalanche deposits from the lake sediment [6,7]. The destructive methods (sieving and LOI) are rather labor-intensive and time consuming. In recent years, the interest in faster and non-destructive methods, such as X-ray computed tomography (CT) scanning, has grown [6]. CT scans provide a 3D reconstruction of the core that can be used to obtain qualitative and quantitative information, such as grain-size parameters and morphology [12]. Together with precise core dating (e.g., varve chronology), X-ray CT from lake sediments can be used to record past snow avalanche events at high temporal (annual) resolution.

In this paper, we investigate the potential of lake sediments to record snow avalanches and to reconstruct the avalanche history. We further explore the impact of transport processes on the significance and the completeness of the snow-avalanche record. Finally, we explore the potential and limitations of the methods used. The study is based on X-ray CT imaging of 27 cores taken from Kenai Lake, Alaska. This technique allows us to examine large clasts, which are hypothesized to be snow-avalanche deposits (dropstones), and the typical varved lake sediments in which the debris was deposited. We revise the varve chronology established by Boes et al. [13] using new absolute ages from cryptotephras present in the varved-sediments, and we compare the annual debris record with satellite images of Kenai Lake to assess the validity of the snow-avalanche hypothesis.

2. Setting

Kenai Lake is a proglacial lake located on the Kenai Peninsula, in south-central Alaska (Figure 1). The Kenai Peninsula can be divided into two regions: the northwest side is the relatively flat Kenai Lowland, and the southeast side is rugged and dominated by the Kenai Mountains where Kenai Lake is situated. The topography of both regions was shaped by Quaternary glacial processes. The Cordilleran Ice Sheet covered most of southcentral Alaska during the Last Glacial Maximum, including the area of Kenai Lake [14]. Kenai Lake was formed by glacial scouring of bedrock and is surrounded by steep mountain slopes.

Kenai Lake is a narrow and zigzag-shaped lake, which is 38 km long and covers a surface area of 56 km^2. The lake is situated 132 m above sea level and has a maximum water depth of 174 m (Figure 1) [14]. The geology of the surrounding mountains was mapped and described [15] as metasedimentary rocks of the Valdez Group (Upper Cretaceous). The mountains south of the lake are higher (1500 m a.s.l.) and steeper (40–45°) than those in the north (600 m a.s.l., 35–40°) (Figure 1). The lake is hydrologically open with the Upper Kenai River, which is the outlet situated on the northwest end of the lake. The Trail

River and Snow River, along the east side of the lake, are the main inlets and sources of water and sediment. A smaller inlet is situated at the north side of the lake [13].

The Kenai Peninsula is situated above the Alaska-Aleutian subduction zone, where numerous active volcanoes are present, and earthquakes occur frequently [16]. These earthquakes and volcanic eruptions can result in the deposition of turbidites and tephras in lakes that can then be used as time markers. In Kenai Lake, the 1964 Great Alaska Earthquake (M_W 9.2) triggered deposition of a turbidite [17], which formed the base of the interval that is studied here. Within this interval, Boes et al. [13] suggested that one of the peaks in the magnetic susceptibility curve, taken along a sediment core, is related to the 1989/1990 Redoubt Volcano eruption.

Kenai Lake sediments are composed of clastic varves, representing a couplet of a coarse-grained spring–summer lamina and a fine-grained winter lamina [13,18]. Boes et al. [13] observed internal laminations in some varves that could be mistaken for a new varve year. These laminations were interpreted to be related to flood events in the late summer or to spring melting of large amounts of snow/ice in a short period.

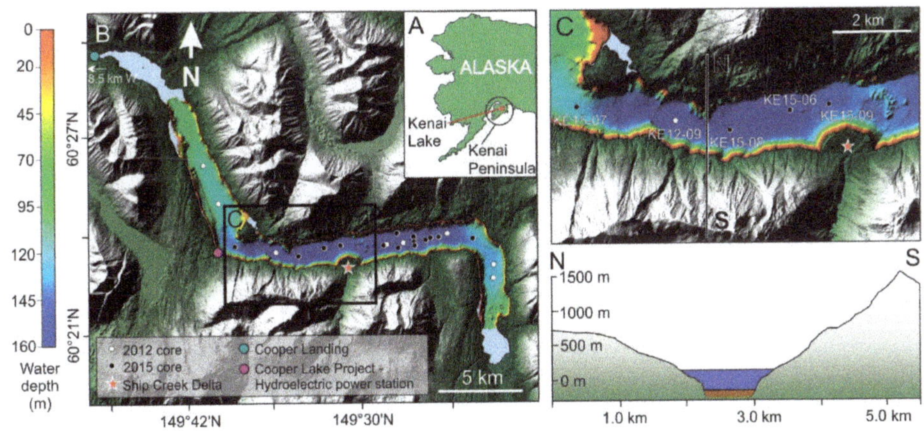

Figure 1. Setting figure of Kenai Lake focussing on the surrounding topography, the bathymetry and the coring locations: (**a**) Map of Alaska showing the location of the study area on Kenai Peninsula. (**b**) Multibeam bathymetry of Kenai Lake on top of a lidar map of the surrounding area [19]. Coring locations of 2012 and 2015 are marked, and villages and deltas are labelled. (**c**) Enlarged western part of the central basin of Kenai Lake with a north-south transect through the central part of the basin.

3. Materials and Methods

3.1. Multibeam Data, Core Collection and Core Analysis

Multibeam bathymetry data for Kenai Lake were collected in the summer of 2015 using an ELAC SeaBeam 1050 system (ELAC Nautik GmbH, Kiel, Germany). This sonar system has two 50 kHz transducer arrays and a 120° swath angle. Roll, pitch, and heave were corrected by an IXSEA Octans 3000 motion sensor (iXBlue, Cedex, France). CTD measurements (conductivity, temperature and depth) at the lake surface were carried out with a Valeport Modus CTD. A CastAway CTD and a Sea and Sun Technology CTD 48M were used for sound velocity profiles, measured several times each day. The echosounder was controlled by ELAC HydroStar software, which also served as an interface for external sensors that acquired hydrographic information. Processing of the bathymetric data was done with HDPedit (ELAC Nautik GmbH, Kiel, Germany)(outlier cleaning), HDPpost (ELAC Nautik GmbH, Kiel, Germany) (correcting of sound velocity profiles) and IVS 3D Fledermaus™ (QPS B.V., Zeist, The Netherlands) (detailed ping editing). The final high-resolution bathymetry map has a 9-m-grid spacing and a vertical accuracy of a few decimeters.

A total of 27 sediment cores were retrieved during the summer from Kenai Lake using a gravity corer with hammering extension; 9 in 2012 and 18 in 2015. After shipping to Belgium, the cores were split at Ghent University, the working halves were described macroscopically, and geophysical parameters were logged at 0.2 cm resolution using the Geotek Multi-Sensor Core Logger (Geotek Ltd, Northamptonshire, United Kingdom) (MSCL). The MSCL is equipped with a Cs gamma source and detector to calculate gamma density, a Konica Minolta CM-2600d spectrophotometer (Konica Minolta Sensing Americas Inc, Ramsey, New Jersey) to analyze colour spectra and a Bartington MS2E point sensor (Barington Instruments Ltd, Oxon, United Kingdom) to measure magnetic susceptibility. Photographs of the split core surfaces were acquired using the line scan camera mounted on the MSCL.

3.2. Core Scanning and Debris Detection

All cores were imaged with a medical X-ray computed tomography scanner (Siemens SOMATOM Definition Flash (Siemens Healthineers, Erlangen, Germany) at the Ghent University Hospital. The scanner was employed at 120 kV, an effective mAs of 200 and a pitch of 0.45. The helical scans were reconstructed using two different protocols. A "soft tissue" reconstruction produced smoother images that have a relatively strong contrast between sediments with low X-ray attenuation. They were used to image the varves in the cores. The "bone" reconstruction produces sharper (but more noisy) images, creating sharp boundaries between objects that have a highly contrasting X-ray attenuation. They were used to image and analyse the medium to coarse sand and gravel in the muddy matrix.

The 2012 cores were split before scanning; while those collected in 2015 were scanned as whole cores. The reconstructed DICOM (Digital Imaging and Communications in Medicine) images have a resolution of 0.15 mm/pixel and a thickness of 0.6 mm and were stacked to create a 3D volume. The step size between consecutive images was 0.6 mm for the 2012 cores, but 0.3 mm for the 2015 cores. Consecutive images of the latter, thus, overlap 50%. The resulting voxel volumes are ~0.014 and ~0.007 mm^3 for 2012 and 2015 cores, respectively. The stacks were imported in VG Studio (2.2) for visualisation and were converted to tiff images in ImageJ (64-bit Java 1.6.0_24; [20]). The tiff images of each core were uploaded in Octopus Analysis (formerly known as Morpho+) [21]. This software allows a 3D-analysis of the cores and the segmentation and isolation of groups of voxels, which represent sand or gravel, based on their specific Hounsfield Units (HU), i.e., characteristic grey values which are related to their specific X-ray attenuation [12,22]. A dual threshold of 2200 to 2250 HU in the 2015 cores and 2350 to 2400 HU in the 2012 cores was used to select and isolate sand or gravel. The higher threshold values for the 2012 cores were determined by comparing the HU histograms and are the result of the drier state of these split cores compared to the closed 2015 cores. For each isolated group of voxels (i.e., sand or gravel grains), the equivalent spherical diameter (ESD) was calculated. The ESD represents the diameter of an irregular shaped object calculated from a sphere with the same volume as the aspheric particle [23]. The minimal grain diameter that theoretically can be distinguished (i.e., ~0.24 mm) is affected by the minimal voxel volume of the CT-images (i.e., ~0.007 mm^3). Following Nesje et al. and Vasskog et al. [7,10], we refer to the particles with an ESD larger than 0.25 mm (medium sand) as debris particles. For each core, we analyzed both the number of particles present and the ESD of each individual isolated particle (Figure 2). The ESD of the debris particles was used to study their average grain size. The number of particles found in each layer represents a minimum, as some grains are too small or are located too close to the core liner, which makes it very difficult or impossible to isolate this group of voxels. The stratigraphic position of each grain was linked to a particular varve. Based on the CT-images and the results of the grain-size analysis of the debris, two cores (i.e., KE15-07 and KE15-08) were selected for sampling of grains of different sizes (Figure 2). These fragments were visually examined to identify the rock type, which can help to assign a possible source region to the debris.

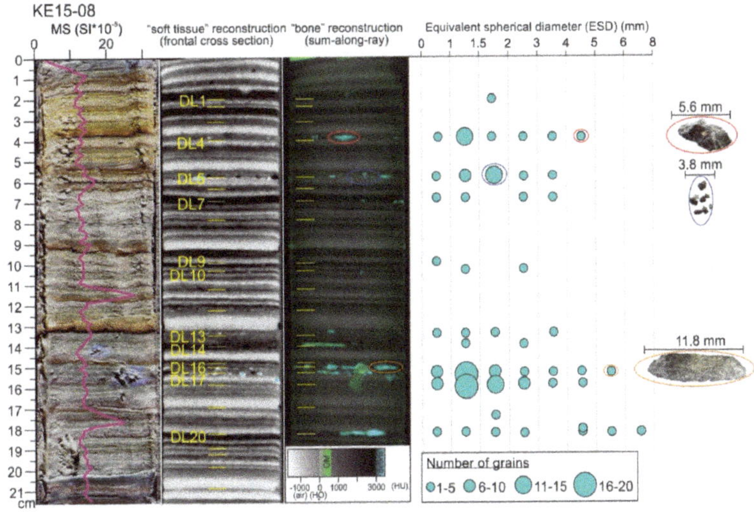

Figure 2. From left to right: Photograph of core KE15-08 with enhanced colour histogram to increase the visibility of the varves with a superimposed record of magnetic susceptibility (MS) in purple. Frontal cross -section of the "soft tissue" reconstructed image. X-ray computed tomography (CT) "sum-along-ray" (equivalent to inverted radiograph) image of the "bone" reconstructed image stack of the same core; laminated (varved) mud is shown in grey scale (intermediate HU), debris in blue (high HU), and organic material in green (low HU). Each layer that contains debris in this core is labelled by DL (debris lamina) and the layer number. The number of grains per equivalent spherical diameter (ESD) class is shown on the right. Pictures of selected grains from three layers.

3.3. Core Correlation, Varve and Tephra Chronology

The cores containing debris were correlated by marker horizons and the sequences of varves present in each core. Varves containing debris were marked as debris laminae (DL) followed by the number of the layer, starting at the sediment–water interface and counting downwards to the top of the turbidite deposited in response to strong shaking during the 1964 earthquake [17]. Dating of the different debris laminae was based on the varve chronology, corroborated by tephra marker beds. The different varves are formed in varve years (VY), which start at the beginning of the melting season (spring) and end at the end of winter [13,18], e.g., VY2014 starts approximately in May 2014 and ends in April 2015.

The varve chronology of the different cores is based on varve counting and chronology by Boes et al. [13]. This original chronology is correlated to historical volcanic eruptions based on newly acquired glass geochemistry data from two cryptotephras indicated by sharp MS peaks (Figure 2). The uncertainty of the varve chronology between marker horizons was calculated using the presence of uncertain varves that were retained for the chronology (a potential "erroneous" varve [17]) and uncertain varves that were not used in the chronology (a potential "missing" varve [17]). From each marker horizon downward, minimum and maximum ages were calculated by summing "erroneous" and "missing" varves, respectively. From each marker horizon upward, minimum and maximum ages were calculated by summing "missing" and "erroneous" varves, respectively. As a result, each lamina receives two minimum and two maximum ages (i.e., one relative to both the nearest marker horizon above and below). From those, the minimum and maximum ages representing the lowest uncertainty were retained.

Cryptotephra samples were treated with 35% hydrogen peroxide to remove organic material and washed through a 20 µm sieve. Glass was separated using LST (lithium heteropolytungstate) heavy liquid at ~2.45 g/cm^3 and mounted in acrylic pucks using SpeciFix-20 epoxy. All analyses were by wavelength-dispersive spectrometry (WDS) on a JEOL 8900 SuperProbe (JEOL USA Inc, Peabody, Massachusetts) at the University of Alberta using 15 KeV, 6 nA current and 5 μm beam. The software 'Probe for EPMA' (Probe Software Inc, Eugene, Oregon) was used to apply time-dependent intensity corrections for Na loss induced by the smaller beam size [24]. Two natural glasses, Lipari obsidian (ID3506) and Old Crow tephra, were analyzed before and after the samples to track calibration and the quality of analyses.

3.4. Satellite Image Analysis

Satellite images were used to visually verify the occurrence of snow avalanches in winters over the period 2004 to 2014. The location and timing of these avalanches were compared to the ages of debris laminae found in the lake sediments. Satellite images from Landsat 5,7,8 /Copernicus (Property Google—DigitalGlobe 15–30 m resolution, 14 April 2004, 17 April 2011), Pleiades (0.5 m resolution, 2 April 2012), QuickBird (0.6 m resolution, 23 March 2014) and WorldView-1 (0.5 m resolution, 28 April 2008) were used to study the frozen conditions of Kenai Lake towards the end of the winter season (March–April). For other years, we were not able to locate satellite images with sufficient resolution.

3.5. Climate Data

The Alaska Climate Research Center provided climate data for the period 1963 to 2013 (for winter temperature and snowfall) collected at two stations located near Kenai Lake. These two stations were Cooper Lake Project (60°23′N, 149°41′W, 182 m above sea level) and Cooper Landing (60°29′N, 149°58′W, 105 m above sea level) (Figure 1).

4. Results

4.1. Avalanches on Satellite Images

We observed snow-avalanche deposits on the five available satellite images over the period 2004 to 2014 along the south shore of the central basin of Kenai Lake (Figures 3 and 4). The majority of these can be found between the Hydroelectric Power Station (HPS) and the Ship Creek Delta (SCD) (Figure 1), while a few are situated east of Ship Creek Delta. We refer to these places as the western part and the eastern part of the central basin, respectively. In the captions of Figures 3 and 4, we describe the distribution and transport processes of observed avalanche deposits based on satellite images.

4.2. Bathymetry

The bathymetry of Kenai Lake can be characterized as a flat basin plain bordered by steep slopes that are typically 30 to 45°, and locally, as steep as 70° (Figure 1). Deltas have average slopes of 20 to 30°. At the foot of major deltas, the hummocky mounds are probably deposits from delta failures.

4.3. Core Sedimentology

Most sediments in Kenai Lake are clastic varves (see 2. Setting; [13]). The varves have a grain-size distribution ranging from clay to silt [13] and form the matrix in which the coarse debris was deposited. The post-1964 sedimentary sequence thickens from the west to the east. The considerably thicker sedimentary sequence in the central and eastern basins (21 to 30 cm) compared to the western basin (11 to 16 cm) (Figure 5) is a result of the river inlets, which are situated in the eastern and central part of the lake. These rivers bring sediment into the lake, which is deposited primarily near the inlets. Within the central basin, the thickness is relatively constant, with slightly thicker (max. 5 cm) sequences on delta-proximal locations.

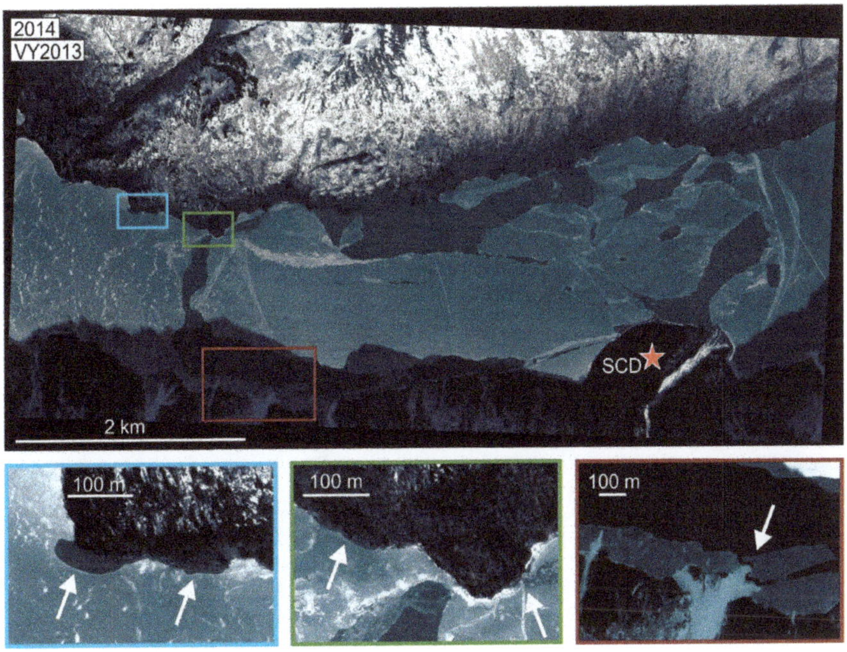

Figure 3. March 23, 2014. Top: Satellite image of the central basin of Kenai Lake. The Ship Creek Delta (SCD) is indicated with a red star as a reference point. Two snow-avalanche deposits occurred in the western part of the central basin with little to no debris visible. Bottom: Magnified images of the observed snow-avalanche deposits and ice melting along the north shore. Melting was followed by the growth of cracks in the ice and finally cracked ice pans that floated around in the lake. Such cracks were also observed along the border of the eastern avalanche deposits.

4.3.1. Debris Laminae

Debris was found in three of the nine cores taken in 2012 (i.e., KE12-07, KE12-08, and KE12-09) and in 11 out of 18 2015-cores (i.e., KE15-01A, KE15-01B, KE15-06, KE15-07, KE15-08, KE15-09, KE15-10, KE15-11, KE15-13, KE15-15, and KE15-17) (Figure 5). In each core, the debris was found in, or just above, the winter lamina of a varve (Figure 2).

Twenty-four debris laminae were identified in the 14 debris-containing cores; most of the debris laminae were found in cores from the southern part of the western basin and in the western part of the central basin (Figures 5 and 6). Core KE15-07, located at the western end of the central basin, records 20 of the 24 debris laminae (Figure 5). The equivalent spherical diameters (ESD) of the debris grains range from 0.35 mm to 12.30 mm. In general, debris consists of very coarse sand to fine gravel. An overview of the physical properties (mean, median, maximum and minimum grain size) of all the grains found in each debris lamina is presented in Table 1.

Figure 4. Left: Satellite images of the central basin of Kenai Lake, from 14 April 2004, 28 April 2008, 17 April 2011, and 2 April 2012. Plotted on the satellite images are the locations of cores, in which debris was found at the top of the varve of the previous varve year. In addition, the median grain size and the number of grains found in that varve of each core are plotted. The Ship Creek Delta (SCD) is indicated with a red star as a reference point. Most of the avalanches (six in 2004, two in 2008, two in 2011, and four in 2012) occurred around the western part of the central basin and two (2004 and 2008) are observed in the eastern part. Avalanches in 2004 and 2008 consist of snow and debris, while those in 2011 and 2012 only show snow. Right: Magnified images of avalanches.

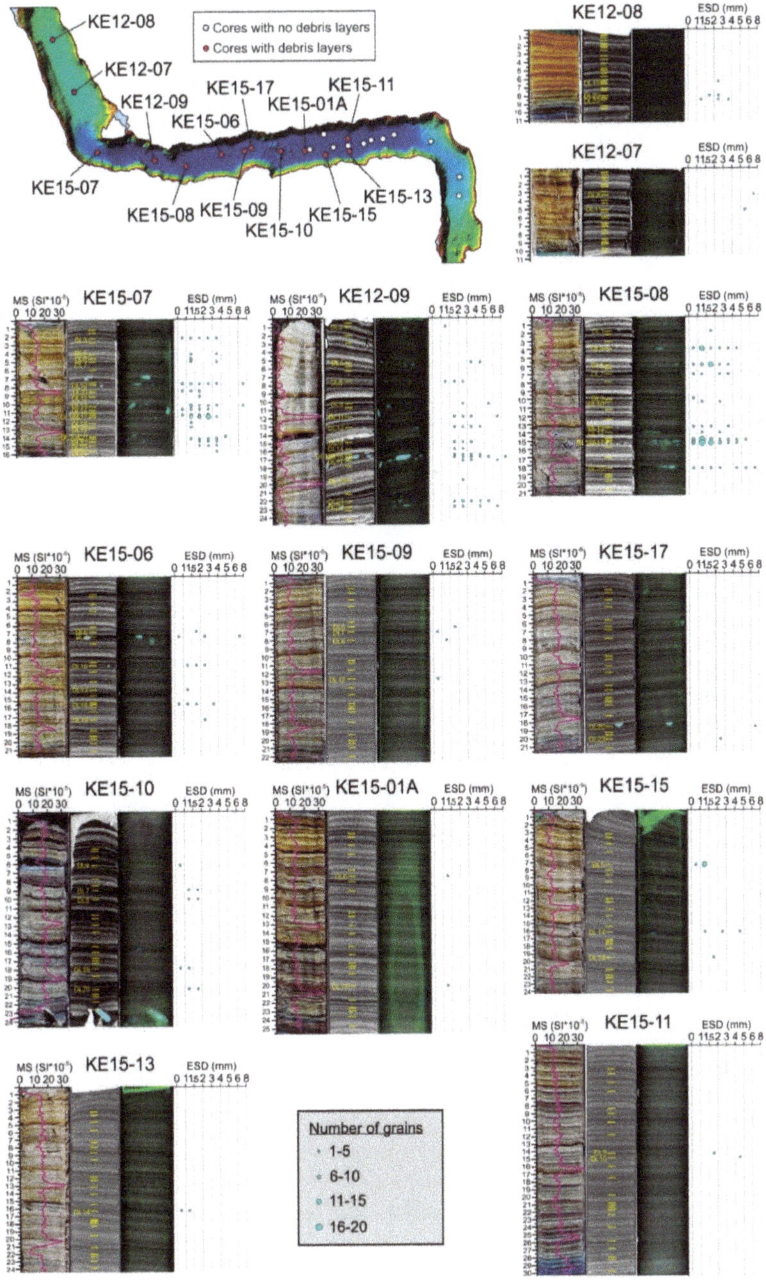

Figure 5. Overview of all cores from Kenai Lake where snow-avalanche debris was found. For each core there is from left to right: photograph of the core with enhanced colour histogram, frontal cross-section of the "soft tissue" reconstructed image. X-ray CT "sum-along-ray" (equivalent to inverted radiograph) image of the "bone" reconstructed image stack and a plot with the number of grains found per debris lamina for different classes of equivalent spherical diameter (ESD).

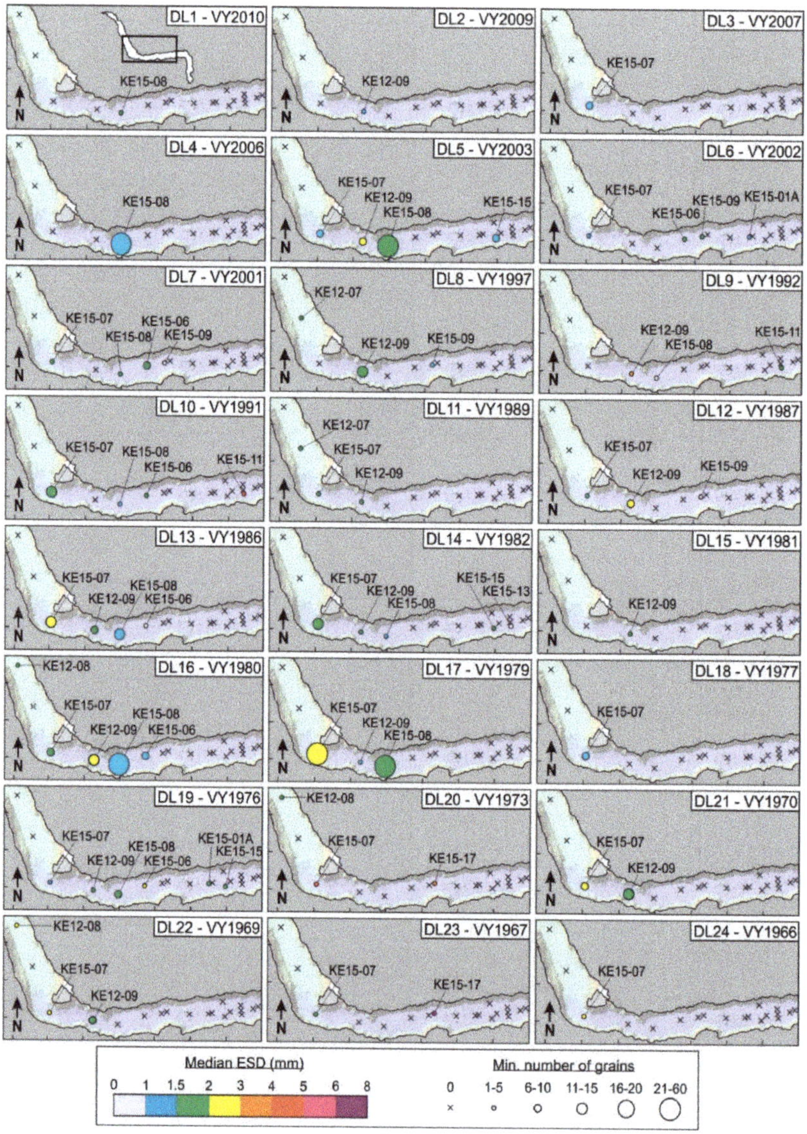

Figure 6. Overview of the 24 debris laminae (DL) located in different cores of Kenai Lake. The cores that contain a certain debris lamina are plotted together with the minimum number of grains found in the lamina and their median equivalent spherical diameter (EDS).

Debris was only sampled in cores KE15-07 and KE15-08. The debris in both cores generally consists of angular, often flattened, grains with a dark gray colour and shiny surface, sometimes accompanied by organic detritus. In debris lamina 4 from core KE15-08 (at 3.5 to 4 cm from the top of the core) one dark grey, angular pebble, 0.6 cm long with a flattened shape and a flaky surface texture was present, which was accompanied by smaller sand grains (<1 mm) (Figure 2). In debris lamina 5 (at 5 to 6 cm from the top) we found sand grains with a diameter up to 1 mm. Whereas most of the grains are angular and dark grey, some grains were more rounded and vary in colour between yellow and dark grey. Organic material, such as wood chips, was found in the matrix, together with debris particles. In debris lamina 16 at a depth of 14.5 to 15.5 cm from the top, sand grains similar to those in debris lamina 4 and 5 were found. In addition to the small grains, we found a 1.2-cm-long, dark grey, angular pebble.

Table 1. Characteristics of debris laminae in the cores, their ages and physical properties of the grains (DL = debris lamina, ESD = equivalent spherical diameter).

Name	Varve Year	Best Age (years AD)	Min Age (years AD)	Max Age (years AD)	Number of Cores in which DL is Present	Total Number of Grains in DL	Mean ESD (mm)	Median ESD (mm)	Max ESD (mm)	Min ESD (mm)
DL1	2010	2011.3	2011.3	2010.3	1	1	1.75	1.75	1.75	1.75
DL2	2009	2010.3	2010.3	2009.3	1	1	1.50	1.50	1.50	1.50
DL3	2007	2008.3	2009.3	2007.3	1	9	1.70	1.68	3.04	0.98
DL4	2006	2007.3	2009.3	2006.3	2	24	1.49	1.17	5.00	0.74
DL5	2003	2004.3	2005.3	2003.3	4	53	1.87	1.52	4.86	0.60
DL6	2002	2003.3	2004.3	2002.3	4	5	1.55	1.57	1.81	1.21
DL7	2001	2002.3	2003.3	2001.3	5	16	2.23	2.15	7.17	0.56
DL8	1997	1998.3	1999.3	1998.3	4	9	3.00	2.51	8.33	1.29
DL9	1992	1993.3	1990.3	1993.3	3	6	2.05	1.29	5.78	0.74
DL10	1991	1992.3	1993.3	1992.3	4	21	2.05	1.69	6.03	0.89
DL11	1989	1990.3	1990.3	1990.3	3	11	2.74	1.90	7.50	0.91
DL12	1987	1988.3	1988.3	1988.3	3	11	2.52	2.06	5.00	0.78
DL13	1986	1987.3	1987.3	1987.3	4	37	1.95	1.70	7.05	0.69
DL14	1982	1983.3	1983.3	1983.3	5	25	1.95	1.61	7.70	0.35
DL15	1981	1982.3	1982.3	1982.3	1	5	2.62	2.51	4.03	1.51
DL16	1980	1981.3	1981.3	1981.3	4	73	2.15	1.51	15.50	0.65
DL17	1979	1980.3	1980.3	1980.3	3	95	1.87	1.64	7.83	0.80
DL18	1977	1978.3	1978.3	1978.3	2	8	1.47	1.49	1.98	1.00
DL19	1976	1977.3	1977.3	1977.3	6	15	2.48	1.62	7.58	0.99
DL20	1973	1974.3	1974.3	1974.3	3	4	4.20	3.60	7.93	1.68
DL21	1970	1971.3	1972.3	1970.3	2	20	2.59	2.23	5.28	1.13
DL22	1969	1970.3	1971.3	1969.3	3	17	2.89	2.64	6.26	1.01
DL23	1967	1968.3	1969.3	1968.3	2	5	2.23	1.90	2.99	1.52
DL24	1966	1967.3	1968.3	1967.3	1	1	2.23	2.23	2.23	2.23

4.3.2. Tephra Geochemistry

The two cryptotephras, which were identified by their magnetic susceptibility peaks, occur throughout the lake (Figure 5). They were sampled in core KE12-05 at a depth of 10.0 to 10.4 cm (UA3196) and 18.7 to 19.0 cm (UA3197). Glass geochemical analyses are shown in Table 2. Both samples have high SiO_2 values. However, sample UA3197 is characterized by relatively high CaO and FeOt, and low K_2O values compared to sample UA3196. All individual glass shard analyses, including standards, are available as supplementary materials (Table S1).

Table 2. Averages and standard deviations (std) of major-element glass geochemistry of tephra samples.

Sample	SiO_2	TiO_2	Al_2O_3	FeOt [1]	MnO	MgO	CaO	Na_2O	K_2O	Cl	H_2Odiff [2]	n [3]
UA3196	77.58	0.27	12.30	1.12	0.06	0.20	0.99	3.85	3.52	0.12	1.27	24
std	1.42	0.06	0.63	0.33	0.02	0.11	0.34	0.31	0.23	0.04	1.26	
UA3197	76.20	0.37	12.57	1.93	0.05	0.42	1.88	4.10	2.21	0.33	1.62	29
std	0.65	0.05	0.27	0.14	0.02	0.06	0.15	0.32	0.36	0.03	1.02	

[1] Fe is presented as total FeO (FeOt); [2] All analyses are normalized to 100% on a water-free basis; [3] n = number of analyses.

5. Discussion

5.1. Chronology

Glass geochemical analyses show that the upper tephra (UA 3196) correlates to the 1989/90 Redoubt Volcano eruption (Figure 7). This tephra is a regional marker and present as a cryptotephra across a large portion of Yukon and Alaska [25]. The second tephra (UA 3197) was, based on its stratigraphic position, presumed to be sourced from the large 1976 eruption of Augustine Volcano with a Volcanic Explosivity Index (VEI) of 4. While we do not have access to reference material from this particular event for comparison, existing data support this correlation. Overall, the glass geochemical characteristics of this sample fall within the variability displayed by historic and pre-historic Augustine tephras, with distinctive traits, such as low K_2O (<2.5 wt%) and higher Cl (~0.25–0.35) contents (e.g., [26,27]). The most similar material are several pumice analyses presented by Roman et al. [27] from the 1986 eruption of Augustine Volcano (Figure 7). The average of their analyses plot with, or very near, UA 3197.

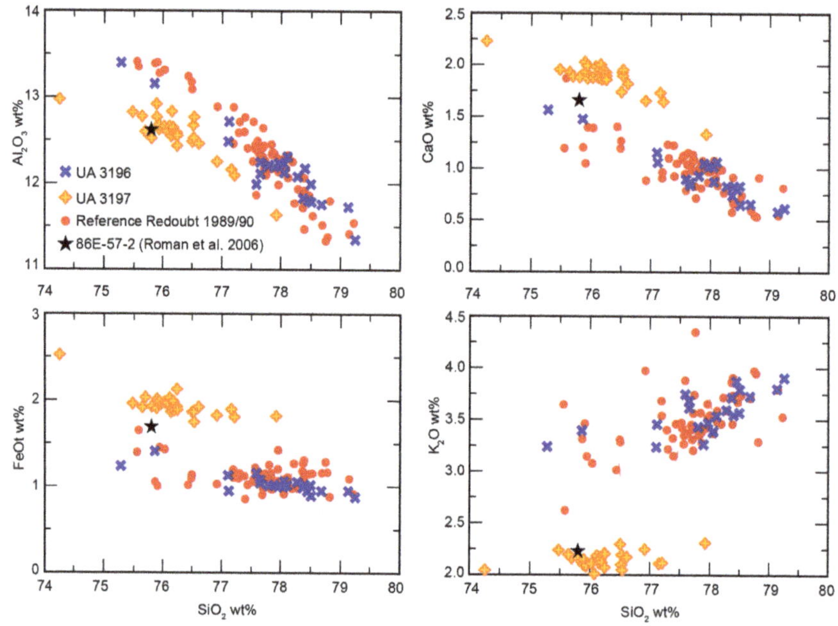

Figure 7. Plots of glass geochemistry of unknowns and reference data. UA 3196 plots with all elements with the reference data for the 1989/90 eruption of Redoubt [25]. Data from Roman et al. [27] of the Augustine eruption are the average of five analyses. Unfortunately, no standard deviation or standards are provided in the paper. However, the analyses for this population identified in the 1986 eruption are similar to UA 3197. All analyses, including standards, are available as supplementary materials (Table S1).

The 1976 eruption of Augustine Volcano had several phases over four months from January until April, including a large event on January 24 where column heights reached ~14,000 m a.s.l., and ashfall was reported on Kenai Peninsula and as far as Anchorage (e.g., [28,29]). The Alaska Volcano Observatory's comprehensive database of eruptions shows that only four other volcanoes had eruptions around this time: Shishalden, Pavlof, Akutan, and the Ukinrek Maars. It is very unlikely that ash from Akutan, Shishalden, and Pavlof would have reached this region; the most proximal of these three volcanoes (Pavlof) is 750 km from Kenai Peninsula, all eruptions had a VEI $\leq$2 with column heights not reaching above 2500 m a.s.l., and no ash was reported beyond the immediate region. Additionally, the eruptions appear to have been largely strombolian involving lava flows, suggesting that products would not have been rhyolitic as erupted at Augustine. The only other large event around this time were the eruptions of the Ukinrek Maars in 1977 (VEI 3) that ejected ash over a wide region, reaching Kodiak Island, south of Kenai Peninsula. However, these were basaltic eruptions (e.g., [30,31]).

We use the two tephra ages to revise the original varve chronology of Kenai Lake by Boes et al. [13] (Figure 8). From the top of the core until the appearance of the first tephra layer (UA3196) the original chronology is unchanged. Only in the lower part of the studied interval, from UA3196 to the 1964-turbidite, the original chronology underwent minor modification. The early 1976 Augustine provenance of tephra UA3197 allows us to attribute the varve with the tephra to varve year 1975. We also know that the position of the 1964-turbidite in the cores was triggered in March 1964 by the M_W 9.2 Great Alaska earthquake [16]. As it likely takes several months, or even longer, to deposit the fine-grained cap of the turbidite [32], the first varve following the turbidite would have been deposited in varve year 1965. The original varve chronology was lacking one varve in between the two absolute time markers (i.e., the 1964 turbidite and the 1976 Augustine tephra), and we, therefore, reinterpret as two varves the multipulsed varve in which debris layer 23 is located (Figure 8).

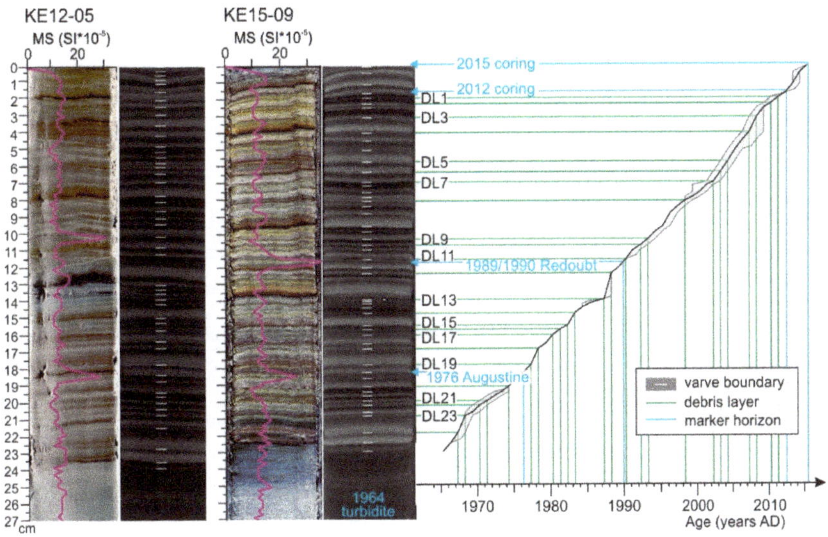

Figure 8. From left to right: Original varve chronology (white lines) of Kenai Lake by Boes et al. [13]. Updated varve chronology based on two cryptotephra layers (lower blue lines). Age–depth model of core KE15-09. For each debris layer (DL; green lines) (y-axis) the age can be found on the x-axis.

5.2. *Evidence for Snow Avalanches as the Origin of Coarse Debris*

The mountains surrounding the lake are the source area of the debris, and the angularity of the fragments indicates that snow avalanches are likely the initial mechanism responsible for the erosion and transport of sand and gravel onto the lake ice. The metasedimentary rocks in the catchment match the rock type of the large clasts from cores KE15-07 and KE15-08. We infer that during spring melting of the lake ice, sediment transported by snow avalanches is then distributed across much of the lake by drifting ice pans.

Satellite images of the Kenai Lake area provide more evidence for snow avalanches as the primary mechanism to bring debris to the lake. Images taken at the end of the winter season (March–April) show snow-avalanche deposits as clearly visible on the frozen lake. However, they do not reach the middle of the lake, where the cores were taken. We explain the deposition of debris in the middle of the lake by the melting processes of the ice in the spring. On the 2014 satellite image (Figure 3), ice along the north coast of the lake starts melting first, and the ice starts cracking across the lake. Cracking of the ice results in ice floes on the lake water, which carry debris initially deposited on the ice near the shore and deliver it to the centre of the lake [9]. The outlet of the lake, the Kenai River, is situated at the northwest end of the lake, and in combination with the predominant easterly wind direction in the months April and May [33], we infer that ice floes would drift westward towards the outlet. Evidence of this westward drift is found in core KE15-07, which contains the most debris laminae of all the cores and is located west of the zone with most snow-avalanche deposits (Figure 5). The other cores with debris were all taken near to, or west of, the source areas, while no debris was found in the cores east of the source areas. Other potential transport mechanisms, such as transport by birds or floating logs, would all be from initially fluvially transported sediment, and the grains would likely be rounded. One would expect a more random distribution of the debris both in time and in space if such a process was responsible for the observed debris. Turbidity currents resulting from onshore rock failure can transport angular grains. We rule out this transport mechanism because the debris particles in the Kenai Lake sediment cores occur typically in the muddy matrix of the varved sediments and not in graded turbidites.

A similar process was proposed to explain the presence of debris in Norwegian [11], Canadian [9], and French lakes [34]. Seierstad et al. [11] observed a dirty snow avalanche, indicating that a significant amount of debris and organic material together with snow and ice was transported by avalanches onto the frozen lake. The process was first described by Luckman [9] who observed it on Canadian lakes in the 1970s. Later, Seierstad et al., Nesje et al., Vasskog et al., and Fouinat et al. [7,10,11,34] explained the distribution of snow-avalanche debris across a lake by the same process. The snow-avalanche debris gets deposited in the spring melting season, at the transition to a new varve year. The coarse particles sink into the fine-grained top of the previous varve. This can be observed in the X-ray CT images (Figures 2 and 5), where all debris particles are situated in the black winter lamina.

5.3. *Snow Avalanche Record*

5.3.1. Variable Avalanche Extent on Satellite Images

The satellite images from Kenai Lake in the month of April over several years show that snow avalanches were deposited on the lake almost every year. However, not every varve contains debris. In years with large snow-avalanche deposits, such as 2004, a debris deposit is present in several cores throughout the lake, whereas smaller avalanche deposits, like in the years 2012 and 2014, are not always represented by debris in one of the sediment cores (Figure 4). The satellite images of these years show that the avalanche deposits are not reaching as far onto the lake ice as they do in other years, and it is possible that during cracking of the ice along the shore, part of the deposit already ends up near the shore [10]. As a result, these snow avalanches are not recorded in the cores taken in the middle of the lake. The 2014 satellite image (Figure 3) shows that ice cracks form around some of the avalanche deposits along the south shore. It is likely that these deposits will not detach from the shoreline, resulting in deposition of the debris on the slopes. Furthermore, depending on which shore of a lake (i.e., shady or sunny—the northern shoreline of Kenai Lake receives more sunlight every day then the south shore, thus lake ice will melt first along the north shore) a snow avalanche is deposited, the debris may or may not get redistributed by drifting ice across the lake.

5.3.2. Natural Factors Affecting the Lacustrine Snow-Avalanche Records

The chance that a snow avalanche is recorded in the lacustrine sediments depends on several factors. The avalanche needs to be large enough and have a sufficiently high velocity so that it spreads far enough onto the ice when it reaches the lake. When an avalanche is too large, it could possibly break the ice when it hits the lake [34], even though this process was not observed on any of the satellite images of Kenai Lake. This process is more likely to happen in the beginning or towards the end of the freezing season as lake ice is still/already thinner and less stable. Only dirty snow avalanches, i.e., avalanches that contain organic material and coarse rock grains, will be recorded in the lake sediments as we can distinguish the coarse avalanche material from the fine lake sediments [9]. In addition to the physical characteristics of the snow avalanche, the melting processes of the lake ice also play an important role in the recording of avalanche deposits in the lake. Depending on where the cracks in the ice form, on wind direction and lake currents [9], the avalanche debris will be deposited at different places across the lake.

Apart from avalanches themselves, the physical characteristics of the lake will influence the record. For example, Lake Lauvitel, studied by Fouinat et al. [6,34] using X-ray CT, is about 130 times smaller than Kenai Lake, which strongly affects the redistribution of avalanche debris. In a smaller lake, such as Lake Lauvitel, there is an increased probability that the snow-avalanche deposits will be spread over most of the lake, thereby increasing the likelihood that a single core will contain a complete record of past avalanches reaching the frozen lake [e.g., 34]. The larger size of Kenai Lake, however, inhibits snow avalanches from reaching the central part of the lake and debris redistribution is, thus, strongly affected by the pathway of drifting ice floes during the melting season. In such large lakes, it is, thus, crucial to conduct a calibration study, such as this one, where multiple short cores are used to map the deposits (Figure 6) and determine the optimal coring locations for a complete record (it should be noted that the cores in this study were not targeted for a snow-avalanche reconstruction). In Kenai Lake, this is the western part of the central basin, in front and to the west of the main avalanche tracks. Cores KE15-07, KE12-09, and KE15-08 contain 18, 14, and 11 out of 24 debris layers, respectively, and combined they contain all 24 recorded debris layers (Figure 5). In the eastern part of the central basin, primarily cores close to the shores contain snow-avalanche debris (e.g., KE15-13 and KE15-15; Figure 5). Such relatively near-shore locations may, thus, contain a more complete avalanche record than cores in the center of the deep basin. Finally, while for a small lake it may be easier to reconstruct a more complete snow avalanche history of a few avalanche tracks, a large lake has the advantage to potentially record avalanches from a larger number of tracks, providing a regional avalanche record.

5.3.3. Comparison of Snow Avalanche and Climate Records

Climate variables, such as mean, maximum and minimum monthly temperature and monthly snowfall (data from the Alaska Climate Research Center), were compared to grain-size distributions and the number of grains from all debris laminae. Baggi et al. [35] suggested that a snowpack is weakened by 1) gradual warming of the snowpack to 0 °C, 2) loss of strength by water infiltration into the snowpack, and 3) overload of the (wet) snowpack by precipitation. These weakening factors of the snowpack could trigger a wet-snow avalanche. In Kenai Lake, we can link the presence of a debris lamina to a varve year where there was snow precipitation in a month with a mean temperature above 0 °C (Figure 9). For example, in 1980 (varve year 1979) a high amount of snow fell in months with a mean temperature above 0 °C and in this varve year, DL17—the debris laminae with most grains of the record—was deposited.

Snow avalanches, specifically slab avalanches, can also initiate from weak bonding of snow above melt–freeze crusts. These crusts typically form during a warmer (melting) period followed by a colder (freezing) period. When fresh snow precipitates on the crust, it will fail to bond to it and can release slab avalanches [36–38]. Our climate data show a link between warmer temperatures in January–February and the presence of debris laminae in that varve year. For example: in 2003 (varve year 2002), when DL6 was deposited, the months of December and January had a mean monthly temperature of −3.4 and −3.2 °C, respectively. In February the mean temperature increased to 0.6 °C and decreased again in March to −4.1 °C. In addition to a drop in temperature, March showed a high amount of snow precipitation. The climate data indicate a melt–freeze crust formed in February-March and that, with increased snow in March, the snowpack became unstable and triggered slab-avalanches. A similar series of events occurred in 2004 (varve year 2003), resulting in DL5, during the year in which most avalanche deposits were observed on the satellite imagery (Figures 3 and 4). We did not identify any relationship among the mean annual winter temperature (Figure 9; yellow plot) or the total annual snowfall (Figure 9; blue plot) and the appearance of debris laminae in lake cores (Figure 9; black plot).

Based on the two mechanisms for avalanche formation, it is possible to link most of the debris laminae to the climate data. However, several other factors can play a role in the triggering of snow avalanches, such as daily temperature and snowfall, wind strength and direction, and snowpack characteristics. The climate data show several years with good snow-avalanche conditions, but during which no debris was found in the lake sediments, as well as years in which debris was found, but weather conditions did not match with snow-avalanche conditions. This may be due to insufficient climate data, coring location, ice conditions, or the redistribution process of the avalanche debris by drifting ice floes.

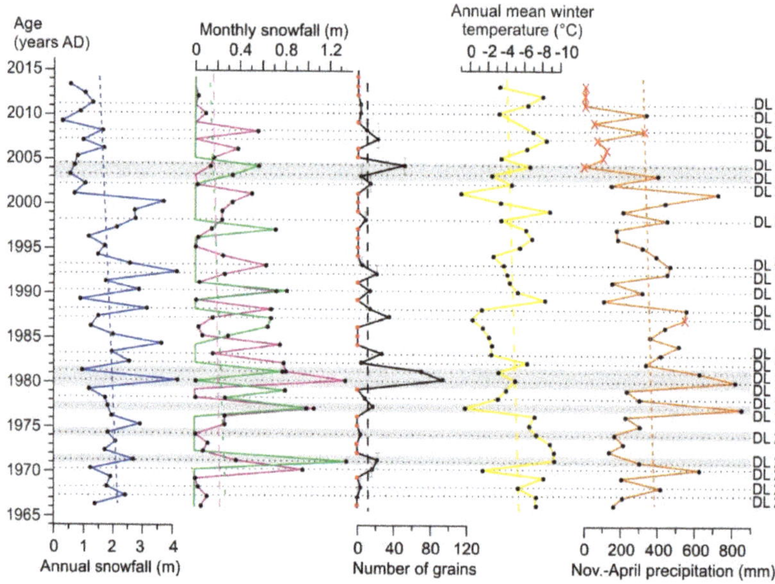

Figure 9. Correlation plot between climate data from Cooper Landing and Cooper Lake Project weather stations (Alaska Climate Research Center) and the snow-avalanche record derived from the sediment cores of Kenai Lake. From left to right: Total annual snowfall. Plots with monthly snowfall when (i) the mean monthly temperature is higher than 0 °C (pink) and (ii) winter temperatures are variable: the mean monthly temperature is rising from November through January, or January and/or February mean temperature is higher than December and/or March average temperature (green). Total number of grains present in each debris lamina over all the cores, red dots indicate that no debris was found in that year. Mean winter temperature measured from December to March. Total precipitation from November to March, red crosses indicate incomplete records. Linear fits of the different plots are indicated by dashed lines. A few examples are highlighted of years with both debris in the cores and high amounts of snowfall during warm months or when the probability of melt-freeze crusts is high due to variable temperatures.

In the 50-year period studied (1965–2015), linear fits of temperature and snow/precipitation plots show decreasing trends. The linear fit of the number of debris laminae, and, thus, the recorded snow avalanches in the lake, also shows a minor decreasing trend. This may be the result of fewer snow avalanches as the amount of snowfall and precipitation decreased during this period.

6. Conclusions

- X-ray CT scans of the cores from Kenai Lake show varved lake sediments and help identify coarse sediment laminae, which we infer as snow-avalanche debris, in a non-destructive methodology. We identified 24 debris laminae in 14 cores, with the coarsest and highest number of grains occurring in the western part of the lake. It is the first time that this technique was used on such a large scale for the identification of snow-avalanche debris.

- We made minor corrections to a previously developed varve chronology using cryptotephras from the 1976 Augustine Volcano eruption and the 1989/90 Redoubt Volcano eruption.
- Based on the rock type of coarse debris and satellite images of avalanche extent over 10 years, we concluded that snow avalanches emanate from the mountains around the lake, and deposit snow and sediment debris onto the frozen lake surface. In spring, the ice breaks up, and ice floes and pans with snow-avalanche deposits drift westward on the lake (forced by river outflow and prevailing wind directions). When the ice melts, debris is deposited on the lake bottom predominantly in locations westward of where the avalanches reached the lake ice. Debris deposition in the varved sediments allows us to date snow-avalanche events with annual resolution.
- Comparison of climate data with the debris laminae record shows that snow avalanches tend to occur in months with a mean positive temperature and increased snow fall. Snow avalanches also occur when there is a sudden temperature rise in winter followed by colder temperatures and snow fall.
- Although the annual debris record is not complete enough to fully reconstruct the regional snow avalanche history, it can be used for decadal reconstructions of past snow-avalanche activity.
- From 1966 to 2014, less snowfall has resulted in a lowering frequency of snow avalanches.

Supplementary Materials: The following are available online at http://www.mdpi.com/2571-550X/2/1/11/s1, Table S1: Individual glass shard analyses of samples and standards.

Author Contributions: Conceptualization, M.V.D.; Methodology, M.V.D. and N.P.; Validation, S.T., M.V.D. and B.J.L.J.; Formal analysis, S.T. and B.J.L.J.; Field work, M.V.D., N.P., E.V. and P.J.H.; Investigation, S.T., M.V.D., B.J.L.J. and T.V.D.; Resources, V.C. and M.D.B.; Data curation, M.V.D.; Writing—original draft preparation, S.T.; Writing—review and editing, M.V.D., N.P., B.J.L.J., P.J.H., V.C. and M.D.B.; Visualization, S.T., M.V.D. and B.J.L.J.; Supervision, M.V.D. and N.P.; Project administration, M.D.B.; Funding acquisition, M.D.B. and M.V.D.

Funding: This research was funded by the Research Foundation—Flanders (FWO), grant number 1523315N.

Acknowledgments: We thank Jasper Moernaut, Michael Strupler, and Adrian Bender for the coring during the summer of 2012, and Koen De Rycker for fieldwork assistance in the summer of 2015. We also want to acknowledge the advice of Katleen Wils on the processing of the Lidar data. We thank Alan Nelson and Benjamin Amann (internal reviewers), 3 anonymous reviewers, and science editor Bernd Zolitschka for providing constructive reviews that improved earlier versions of this paper. Any use of trade, product, or firm names is for descriptive purposes only and does not imply endorsement by the U.S. Government.

Conflicts of Interest: The authors declare no conflict of interest.

References

1. Colorado Avalanche Information Center. Available online: http://avalanche.state.co.us/accidents/statistics-and-reporting/ (accessed on 17 October 2018).
2. Techel, F.; Jarry, F.; Kronthaler, G.; Mitterer, S.; Nairs, P.; Pavšek, M.; Valt, M.; Farms, G. Avalanche fatalities in the European Alps: Long-term trends and statistics. *Geogr. Helv.* **2016**, *71*, 147–159. [CrossRef]
3. Stethem, C.; Jamieson, B.; Schaerer, P. Snow avalanche hazard in Canada—A review. *Nat. Hazards* **2003**, *28*, 487–515. [CrossRef]
4. Ballesteros-Cánovas, J.A.; Trappmann, D.; Madrigal-González, J.; Eckert, N.; Stoffel, M. Climate warming enhances snow avalanche risk in the Western Himalayas. *Proc. Natl. Acad. Sci. USA* **2018**, *115*, 3410–3415. [CrossRef] [PubMed]
5. Butler, D.R.; Malanson, G.P. A History of High-Magnitude Snow Avalanches, Southern Glacier National Park, Montana, USA. *Mt. Res. Dev.* **1985**, *5*, 175–182. [CrossRef]
6. Fouinat, L.; Sabatier, P.; Poulenard, J.; Reyss, J.L.; Montet, X.; Arnaud, F. A new CT scan methodology to characterize a small aggregation gravel clast contained in a soft sediment matrix. *Earth Surf. Dyn.* **2017**, *5*, 199–209. [CrossRef]
7. Nesje, A.; Bakke, J.; Dahl, S.O.; Lie, Ø.; Bøe, A.G. A continuous, high-resolution 8500-yr snow-avalanche record from western Norway. *Holocene* **2007**, *17*, 269–277. [CrossRef]

8. Butler, D.R.; Sawyer, C.F. Dendrogeomorphology and high-magnitude snow avalanches: A review and case study. *Nat. Hazards Earth Syst. Sci.* **2008**, *8*, 303–309. [CrossRef]
9. Luckman, B.H. Drop Stones Resulting from Snow-Avalanche Deposition of Lake Ice. *J. Glaciol.* **1975**, *14*, 1–3. [CrossRef]
10. Vasskog, K.; Nesje, A.; Støren, E.N.; Waldmann, N.; Chapron, E.; Ariztegui, D. A holocene record of snow-avalanche and flood activity reconstructed from a lacustrine sedimentary sequence in Oldevatnet, western Norway. *Holocene* **2011**, *21*, 597–614. [CrossRef]
11. Seierstad, J.; Nesje, A.; Dahl, S.O.; Simonsen, J.R. Holocene glacier fluctuations of Grovabreen and Holocene snow-avalanche activity reconstructed from lake sediments in Grøningstølsvatnet, western Norway. *Holocene* **2002**, *12*, 211–222. [CrossRef]
12. Cnudde, V.; Boone, M.N. High-resolution X-ray computed tomography in geosciences: A review of the current technology and applications. *Earth-Sci. Rev.* **2013**, *123*, 1–17. [CrossRef]
13. Boes, E.; Van Daele, M.; Moernaut, J.; Schmidt, S.; Jensen, B.J.L.; Praet, N.; Kaufman, D.; Haeussler, P.; Loso, M.G.; De Batist, M. Varve formation during the past three centuries in three large proglacial lakes in south-central Alaska. *Bull. Geol. Soc. Am.* **2018**, *130*, 757–774. [CrossRef]
14. Reger, R.D.; Sturmann, A.G.; Berg, E.E.; Burns, P.A.C. *A Guide to the Late Quaternary History of Northern and Western Kenai Peninsula, Alaska*; Guidebook 8; Division of Geological & Geophysical Surveys: Fairbanks, AK, USA, 2007; pp. 1–120.
15. Bradley, D.C.; Kusky, T.M.; Haeussler, P.; Goldfarb, R.; Miller, M.; Dumoulin, J.A.; Nelson, S.W.; Karl, S. Geologic signature of early Tertiary ridge subduction in Alaska. *Spec. Pap. Soc. Am.* **2003**, 19–50. [CrossRef]
16. Carver, G.; Plafker, G. Paloseismicity and neotectonics of the Aleutian subduction zone—An overview. In *Acive Tectonics Seism. Potential Alaska*; American Geophysical Union: Washington, DC, USA, 2008; pp. 43–63.
17. Praet, N.; Moernaut, J.; Van Daele, M.; Boes, E.; Haeussler, P.J.; Strupler, M.; Schmidt, S.; Loso, M.G.; De Batist, M. Paleoseismic potential of sublacustrine landslide records in a high-seismicity setting (south-central Alaska). *Mar. Geol.* **2017**, *384*, 103–119. [CrossRef]
18. Zolitschka, B.; Francus, P.; Ojala, A.E.K.; Schimmelmann, A. Varves in lake sediments—A review. *Quat. Sci. Rev.* **2015**, *117*, 1–41. [CrossRef]
19. DGGS Elevation Portal. Available online: https://elevation.alaska.gov (accessed on 17 July 2018).
20. Schneider, C.A.; Rasband, W.S.; Eliceiri, K.W. NIH Image to ImageJ: 25 years of image analysis. *Nat. Methods* **2012**, *9*, 671–675. [CrossRef] [PubMed]
21. Witte, D.; Brabant, L.; Vlassenbroeck, J.; De Witte, Y.; Cnudde, V.; Boone, M.N.; Dewanckele, J.; Van Hoorebeke, L. Three-Dimensional Analysis of High-Resolution X-ray Computed Tomography Data with Morpho+. *Microsc. Microanal.* **2011**, *17*, 252–263. [CrossRef]
22. Kamalian, S.; Lev, M.H.; Gupta, R. Computed tomography imaging and angiography—Principles. *Handb. Clin. Neurol.* **2016**, *135*, 3–20. [CrossRef] [PubMed]
23. Jennings, B.R.; Parslow, K. Particle Size Measurement: The Equivalent Spherical Diameter. *Proc. R. Soc. Lond. A Math. Phys. Sci.* **1988**, *419*, 137–149. [CrossRef]
24. Donovan, J.J.; Kremser, D.; Fournelle, J.H.; Goemann, K. *Probe for EPMA: Acquisition, Automation and Analysis*; Probe Software, Inc.: Eugene, OR, USA, 2015.
25. Davies, L.J.; Jensen, B.J.L.; Froese, D.G.; Wallace, K.L. Late Pleistocene and Holocene tephrostratigraphy of interior Alaska and Yukon: Key beds and chronologies over the past 30,000 years. *Quat. Sci. Rev.* **2016**, *146*, 28–53. [CrossRef]
26. Blockley, S.P.E.; Edwards, K.J.; Schofield, J.E.; Pyne-O'Donnell, S.D.F.; Jensen, B.J.L.; Matthews, I.P.; Cook, G.T.; Wallace, K.L.; Froese, D. First evidence of cryptotephra in palaeoenvironmental records associated with Norse occupation sites in Greenland. *Quat. Geochronol.* **2015**, *27*, 145–157. [CrossRef]
27. Roman, D.C.; Cashman, K.V.; Gardner, C.A.; Wallace, P.J.; Donovan, J.J. Storage and interaction of compositionally heterogeneous magmas from the 1986 eruption of Augustine Volcano, Alaska. *Bull. Volcanol.* **2006**, *68*, 240–254. [CrossRef]
28. Shackelford, D.C. Augustine: In Annual report of the world volcanic eruptions in 1976 with supplements to the previous issues. *Bull. Volcan. Erupt.* **1978**, *16*, 53–55.
29. Kamata, H.; Johnston, D.A.; Waitt, R.B. Stratigraphy, chronology, and character of the 1976 pyroclastic eruption of Augustine volcano, Alaska. *Bull. Volcanol.* **1991**, *53*, 407–419. [CrossRef]

30. Kienle, J.; Kyle, P.R.; Self, S.; Motyka, R.J.; Lorenz, V. Ukinrek Maars, Alaska, I. April 1977 eruption sequence, petrology and tectonic setting. *J. Volcanol. Geotherm. Res.* **1980**, *7*, 11–37. [CrossRef]
31. Self, S.; Kienle, J.; Huot, J.P. Ukinrek Maars, Alaska, II. Deposits and formation of the 1977 craters. *J. Volcanol. Geotherm. Res.* **1980**, *7*, 39–65. [CrossRef]
32. Van Daele, M.; Meyer, I.; Moernaut, J.; De Decker, S.; Verschuren, D.; De Batist, M. A revised classification and terminology for stacked and amalgamated turbidites in environments dominated by (hemi)pelagic sedimentation. *Sediment. Geol.* **2017**, *357*, 72–82. [CrossRef]
33. NOAA-National Weather Service. Available online: https://w2.weather.gov/climate/index.php?wfo=pafc (accessed on 17 August 2018).
34. Fouinat, L.; Sabatier, P.; David, F.; Montet, X.; Schoeneich, P.; Chaumillon, E.; Bachiller-Jareno, N.; Arnaud, F. Wet avalanches: Long-term evolution in the Western Alps under climate and human forcing. *Clim. Past* **2018**, *14*, 1299–1313. [CrossRef]
35. Baggi, S.; Schweizer, J. Characteristics of wet-snow avalanche activity: 20 years of observations from a high alpine valley (Dischma, Switzerland). *Nat. Hazards* **2009**, *50*, 97–108. [CrossRef]
36. Seligman, G. *Snow Structure and Ski Fields*; International Glaciological Society: Cambridge, UK, 1936.
37. Jamieson, B. Formation of Refrozen Snowpack Layers. *Rev. Geophys.* **2006**, *44*, 1–15. [CrossRef]
38. McClung, D.M.; Schaerer, P.A. *The Avalanche Handbook*, 3rd ed.; The Mountaineers Books: Seattle, WA, USA, 2006.

quaternary

MDPI

Article

Modern Analogue Approach Applied to High-Resolution Varved Sediments—A Synthesis for Lake Montcortès (Central Pyrenees)

Teresa Vegas-Vilarrúbia [1,*], Valentí Rull [2], Maria del Carmen Trapote [1,2], Min Cao [3], Antoni Rosell-Melé [3,4], Teresa Buchaca [5], Joan Gomà [1], Pilar López [1], Javier Sigró [6], Elisabet Safont [1], Núria Cañellas [2], Sandra Garcés-Pastor [1], Santiago Giralt [2], Juan Pablo Corella [7] and Núria Pérez-Zanón [6]

1 Faculty of Biology, Universitat de Barcelona, Av. Diagonal 643, 08028 Barcelona, Spain; mctrapote84@gmail.com (M.d.C.T.); jgoma@ub.edu (J.G.); marilopez@ub.edu (P.L.); esafont@ub.edu (E.S.); sgarcespastor@gmail.com (S.G.-P.)
2 Institute of Earth Sciences Jaume Almera (ICTJA-CSIC), C. Lluis Sole i Sabarís s/n, 08028 Barcelona, Spain; vrull@ictja.csic.es (V.R.); nuriacatcb@gmail.com (N.C.); sgiralt@ictja.csic.es (S.G.)
3 Institut de Ciència i Tecnologia Ambientals (ICTA-UAB), Universitat Autònoma de Barcelona, Bellaterra, 08193 Catalonia, Spain; smilecaomina@hotmail.com (M.C.); antoni.rosell@uab.cat (A.R.-M.)
4 Institució Catalana de Recerca i Estudis Avançats (ICREA), Barcelona, 08010 Catalonia, Spain
5 Centre for Advanced Studies of Blanes (CEAB-CSIC), Accés a la Cala St. Francesc 14, 17300 Blanes, Spain; buch@ceab.csic.es
6 Center for Climate Change (C3), Universitat Rovira i Virgili, C/ Joanot Martorell 15, Vila-seca, 43500 Tarragona, Spain; javier.sigro@urv.cat (J.S.); nuria.perez@bsc.es (N.P.-Z.)
7 Universite Grenoble Alpes, CNRS, IRD, Grenoble INP, IGE, 38000 Grenoble, France; pcorella@iqfe.csic.es
* Correspondence: tvegas@ub.edu

Academic Editors: Bernd Zolitschka and Matthew Peros
Received: 21 May 2019; Accepted: 19 December 2019; Published: 2 January 2020

check for updates

Abstract: In Quaternary paleosciences, the rationale behind analogical inference presupposes that former processes can be explained by causes operating now, although their intensity and rates can vary through time. In this paper we synthesised the results of different modern analogue studies performed in a varved lake. We discuss their potential value to obtain best results from high resolution past records. Different biogeochemical contemporary processes revealed seasonality and year-to-year variability, e.g., calcite precipitation, lake oxygenation, production and deposition of pollen and phytoplankton growth. Fingerprints of the first two of these processes were clearly evidenced in the varve-sublayers and allow understanding related to past events. Pollen studies suggested the possibility of identifying and characterizing seasonal layers even in the absence of varves. Marker pigments in the water column were tightly associated with phytoplankton groups living today; most of them were identified in the sediment record as well. We observed that 50% of these marker pigments were destroyed between deposition and permanent burying. In another study, seasonality in the production/distribution of branched glycerol dialkyl glycerol tetraethers (brGDGTs) and derived temperature estimates were investigated in catchment soils and particles settling in the lake. The signatures of brGDGTs in depositional environments mainly were representative of stable conditions of soils in the catchment that last over decades; no brGDGTs seemed to be produced within the lake. The main contribution of this review is to show the advantages and limitations of a multiproxy modern-analogue approach in Lake Montcortès as a case study and proposing new working hypotheses for future research.

Keywords: endogenic varves; calcite precipitation; pollen traps; meromixis; freshwater GDGTs; pigments; long-term ecology

1. Introduction

Analogy is a powerful form of reasoning that allows two domains to be considered as similar, based on their relationships [1–3]. The analogue approach permits the extension of knowledge of a base domain to a target past domain by virtue of their expected similarity and by means of analogical inference. This process involves reproducing structures and processes from the base to the target in which missing information is generated. Nonetheless, constraints must be placed on what information is to be carried over [4], in order to obtain inferences that are as realistic as possible. In Quaternary paleosciences, the rationale behind analogue inferences is based on postulating that past natural changes can be explained by causes operating today (principle of uniformitarianism), but that intensity and rates at which processes take place vary through time [5]. On the other hand, reconstruction and derived inferences for any proxy (natural recorders of variability) demand a thorough comprehension of the spatial and temporal frame in which processes and environmental conditions leading to the existence and persistence of each particular proxy make sense.

Examples of modern analogue studies of proxies and their applications are widespread. In paleoecology, fossil assemblages (pollen, diatoms, pigments, ostracods etc.) are examined for coincidences with modern assemblages, and the direction and rate at which communities vary through space and time can be measured quantitatively by a wide array of numerical approaches and techniques [6–8]. For instance, in the Peruvian Andes, the pollen/altitude relationship in a forest transect was studied by means of correspondence analysis, and the good statistical correlation between altitude and the first component allowed building a transfer function with statistical reliability to infer altitude from pollen data [9]. In another case, a surface-sediment survey of pigments in 100 lakes (Northern Sweden) was combined with a reconstruction of sedimentary pigments from Lake Seukokjaure. Light regime and terrestrial supplies of ^{13}C-depleted carbon turned out to be important controls of primary production, whereas depth, elevation and conductivity discriminated between different phototrophic communities. The application of these results to the sediment record demonstrated the importance of terrestrial vegetation for the development of the lake, and of specific factors to control algal community composition, across the Holocene [10]. The modern analogue approach can be used to build models for climate change reconstruction. Central to these models is the development of transfer functions that, by definition, are "empirically derived equations for calculating quantitative estimates of past atmospheric or oceanic conditions from paleontological data" [11]. Transfer functions have been widely applied during the last decades. For example, in Patagonia, chironomids of surface sediment of 63 lakes were used to model the atmospheric annual average temperature. The good correlation between estimated and measured temperatures allowed applying the model to the chironomid record of Lake Potrok Aike. The reconstructed trends of temperature were consistent with other quantitative records of the Southern Hemisphere [12]. In the Pyrenees, present chrysophyte cysts distribution is influenced by altitude, and a transfer function was developed to assess the local "altitude anomaly" on a lake site throughout time. A warming trend was detected showing sub-millennial climatic variability from the early Holocene to 4 kiloyears (kyr) before present (BP) [13]. Sometimes, finding modern analogues is not straightforward. For example, a modern analogue study of the Otter lake helped understand why and how siderite ($FeCO_3$) accumulated only in varved sediments older than 1200 calibrated years, even though limnological data did not evidence siderite precipitation. The authors combined modern data with siderite time series, abundances of major element in sediment and experimental results on siderite precipitation. They showed that the lake system alternated between ferruginous, meromictic phases with intense precipitation of siderite and holomictic phases with very low siderite precipitation [14].

Temporal resolution that can be achieved with non-laminated sediment is often constrained to more than decadal resolution. Reconstructions of environmental variability at high resolution scales using varved sediments have provided an unprecedented level of detail that needs to be properly extracted and interpreted [15]. This implies knowledge of fine-tuned mechanisms that need to be observed today at the same temporal resolution as they occur in past records [16]. Lakes with varved sediments are especially well suited for paleoecological and paleoclimatic research demanding a high level of detail, as it is possible to downscale to annual and seasonal resolution [15,17]. The interpretative power of such high-resolution paleoenvironmental reconstructions depends on the availability of suitable modern analogues with the same temporal resolution [16]. With the advent of new techniques and proxies that allow acquiring an increased level of detail, modern analogue studies are gaining importance, a fact we want to highlight with this paper.

We have taken advantage of the varved sediments of the Mediterranean Lake Montcortès and conducted several independent modern analogue studies. Our aim was to document the potential of the modern analogue approach not only by determining how similar modern and past communities are, but also by investigating different contemporary biological and physico-chemical processes that are expected to also have occurred in the past on annual and seasonal time scales. We also explored which factors and relationships between modern and subfossil proxies can limit or hinder the applicability of the modern analogue approach representing a potential source of error in analogue analysis. Such errors should be foreseen and minimised [6]. In this context, we examined how the different modern analogue approaches can help extract important cause-effect relationships at seasonal, annual and subdecadal time scale. We discussed, evaluated and synthesised the published results of four modern analogue studies carried out for Lake Montcortès:

(1) Relationships between calcite precipitation, environmental variables and primary producers and their influences on the thickness and patterns of calcareous varve sublayers were examined as high-resolution paleoclimate indicators [18].
(2) Seasonal pollen sedimentation was compared with local meteorological variables to determine whether contemporary seasonal deposition is consistent with the two layers of the varve pattern archived in the sediments [19].
(3) Contemporary shifts in dissolved O_2, Fe, Mn and anaerobic phototrophic bacteria markers were assessed, in order to investigate whether mixing and oxygenation dynamics Lake Montcortès have been the same over the last five centuries [20].
(4) Distribution of branched glycerol dialkyl glycerol tetraethers (GDGTs) in settling particulate matter and catchment soils was analysed in order to determine whether branched glycerol dialkyl glycerol tetraethers (brGDGTs) are seasonally biased in environments where annual brGDGTs production may not be constant [21].

Additionally, we provided and discussed the first results from ongoing research that aimed at relating contemporary phytoplankton taxa in Lake Montcortès to their corresponding marker pigments, and then to subfossil pigment records at subdecadal time scales.

2. Environmental Settings

Lake Montcortès (Figure 1) belongs to the southern flank of the Central Pyrenees (Catalonia, Spain) at 42°19′N, 0°59′E and 1027 m altitude (Figure 1). This karstic lake located on carbonated and evaporitic Mesozoic substrate is chiefly characterised by Triassic limestones, marls and evaporites, as well as Oligocene-carbonate conglomerates [22,23]. Climatic data collected at La Pobla de Segur meteorological station show that the annual average air temperature is 12.8 °C (2.9 °C in January to 23.2 °C in July). Total annual precipitation is 669 mm (33.4 mm in February and 88.4 mm in May). Maximum and minimum annual air temperatures were 41 and −20 °C, respectively. The vegetation of the catchment area is evergreen with deciduous oak forests (*Quercus rotundifolia* and *Q. pubescens*), conifer forests (*Pinus nigra*), pastures and crops. The vegetation belt at the shore is dominated by hygrophyte communities of *Typha domingensis* or *Cladium mariscus*, reed beds of *Phragmites australis* and communities of *Carex riparia*. Rush formations and grasslands grow on occasionally flooded soils [24]. The lake surface area is 0.14 km^2 and its maximum water depth 32 m [25]. Only an ephemeral inlet exists, but water is drained by an outlet located at the northern shore. The lake's sedimentary record shows thin, well preserved biogenic varves formed during the Late Holocene. The varve structure alternates between sublayers of endogenic calcite and organic detritus with additional detrital layers and turbidites embedded in the varve succession [22,26,27].

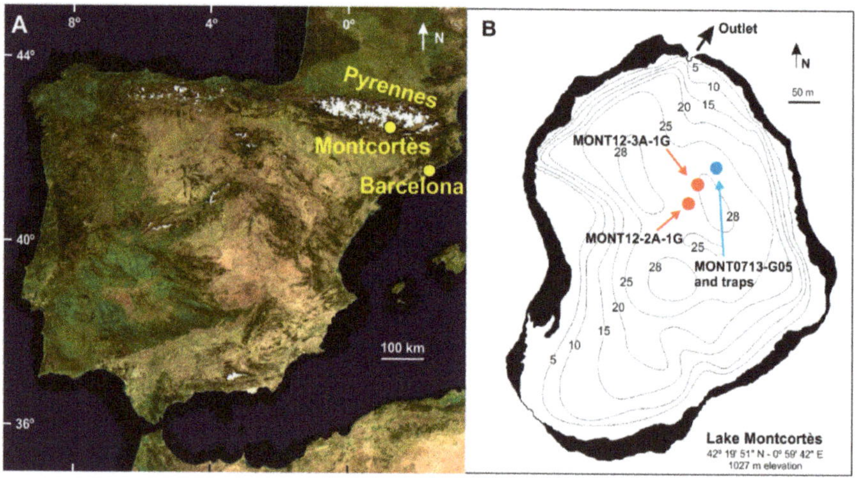

Figure 1. (**A**) Location of Lake Montcortès in the Pyrenean mountain range (NE Iberian Peninsula). (**B**) Bathymetric map with position of the cores and the traps.

3. Materials and Methods

Here, we provided a summary of the main field and laboratory methods used. More detailed information can be found in corresponding publications and literature therein [18–20].

3.1. Monitoring and Coring

The limnological characteristics of Lake Montcortès were assessed monthly from October 2013 to October 2015. Detailed vertical profiles of temperature (T), dissolved oxygen (DO), electric conductivity (EC) and pH were obtained using a [28,29] multiparametric datasonde Hydrolab DS5. Water samples were obtained with an adapted submersible bilge pump at three depths in the epi-, meta- and hypolimnion for alkalinity, cations, total phosphorus (TP), total nitrogen (TN), suspended solids and molecular marker analyses (pigments and brGDGTs), and at two depths (epi- and metalimnion) for phytoplankton microscopic examination. Between 1.5 and 3 L of water was filtered through Whatman GF/F glass-fibre filters for pigment analysis. These filters were protected with aluminium foil and frozen to prevent degradation until laboratory analysis.

A set of sediment traps was deployed at 20 m water depth from a floating platform located at the deepest part of the lake, coinciding with the site where sediment coring was performed. Total suspended solids, calcite crystals [18], pollen and spores [19], marker pigments and settling particulate matter for brGDGT extractions [20] were collected in these traps, with monthly or quarterly (three months) frequency depending on the variable of interest. Additionally, replicated topsoil samples from a depth of 2–5 cm were collected monthly for brGDGT analysis at three sites within a 10 m radius in the catchment area. These samples were combined to reduce soil heterogeneity.

Two UWITEC gravity cores (MON12-3A-1G, 78 cm length, and MON12-2A-1G, 106 cm length) were retrieved at the deepest area of the lake basin (30 m depth) at the same location of the sediment trap (Figure 1) [27]. The varve chronology was performed by double counting in 14 overlapping thin sections. Only 1% of varves were interpolated due to poor varve preservation [27]. Two independent dating methods using AMS ^{14}C and ^{210}Pb radiometric dating techniques were previously achieved in the Lake Montcortès sedimentary sequence [22,26,27]. The agreement between these independent chronologies supports the robustness of the varve chronology.

3.2. Laboratory Analyses

The calcite saturation index (Ω) was estimated from carbonate and calcium concentrations. Ca^{2+} was measured with inductively coupled plasma atomic emission spectroscopy, and $CO_3{}^{2-}$ was calculated from alkalinity, pH and temperature with the CO2SYS_XLS v 2.1 program [28]. To confirm the presence of calcite crystals and to assess seasonal variations of their size and shape, trapped material from traps was analysed by scanning electron microscopy. Total suspended solids retained in the traps were analysed for total particulate carbon (TPC) and total particulate organic carbon (POC) with an elemental organic analyser, Thermo EA 1108.

Diatom specimens of quarterly sediment traps were prepared, identified and counted using standard methods [29]. A minimum of 300 valves were counted per sample; identification was carried out following Krammer and Lange-Bertalot [30,31]. Epilimnetic and metalimnetic phytoplankton was settled using sedimentation chambers; quantification was performed applying Utermöhl's technique for counting combined with biovolume conversion as $mm^3\ L^{-1}$. Analysis of samples for pollen analysis followed standard methods [32] after addition of exotic *Lycopodium* spores. Pollen sum included all pollen types except the superabundant *Pinus* and *Quercus*. See the original palynological work for more detailed information on chemical processing, counting and plotting/statistical procedures [11]. As in earlier studies [33–35], identification of pollen was based on our own general reference collection for the Pyrenees, enriched after detailed floristic and vegetation studies around Lake Montcortès area [24].

Geochemical data to trace high resolution oxic/anoxic events of the hypoliminon were obtained on core MON12-3A-1K by X-ray fluorescence (XRF) analysis (2000 A, 10–30 kV and 20–50 s measuring time) (0.2 mm resolution) using an XRF core-scanner (Avaatech Analytical X-Ray Technology, Netherlands). We obtained sub-annual resolution time series for Fe, Mn, Ca, Si, S, Ti and Br. The former five elements were normalised to Ti. The Fe/Mn ratio was used as the main redox tracer, together with marker pigments of sulphur bacteria, such as okenone (oken) and isorenieratene (isore), which gave information at the subdecadal level.

A second core (MONT-0713-G05) was used for pigment analyses; sediment samples were obtained at sampling intervals of 5 mm, reaching subdecadal resolution, the frozen sediment samples were freeze dried before analysis. Pigment analyses were also performed on epi-, meta- and hypolimnetic water samples and sediment trap material collected in Whatman GF/F glass fibre filters. Marker pigments were extracted from frozen filters or freeze-dried sediment in 5 ml of 90% acetone. The extract was centrifuged, filtered and analysed with ultra-high-performance-liquid-chromatography following the method of [36] with some modifications. Pigments were identified by evaluating their retention times and absorption spectra against a library based on algae and photosynthetic bacterial cultures [37].

To isolate GDGTs, soil samples and filters from sediment traps were extracted monthly, evaporated and then fractionated over an aminopropyl silica column. The fractions containing the lipids were dried, re-dissolved and filtered. brGDGTs were analysed by high-performance liquid chromatography coupled with mass spectrometers and detected in selected ion monitoring modes; prior to extraction, a synthetic tetraether lipid with a structure typical of neutral archaeal membrane lipids was added to the sample. Further analyses in soils were performed for water content and total organic carbon (TOC).

4. Results

4.1. Biogeochemical Processes Involved in Varve Formation: Where, When, and How?

Processes driving varve formation are remarkably variable and site-specific. Consequently, detailed knowledge of the local processes that promote and modulate particle flux dynamics to the sediment for each site becomes essential. This knowledge is a pre-requisite for proper paleoenvironmental and paleoclimatic interpretations of the sedimentary archive [17,38] and for building reliable varve-based chronologies. An independent absolute varve chronology was obtained from the composite sedimentary sequence of both sediment cores extending from 2012 back to Common Era (CE) 1347 [27]. The varves of Lake Montcortès are formed by couplets of white calcite and brownish organic layers. The white layer was assumed to have deposited during spring/summer and the dark layer in fall/winter [17,27]. Biogenic varve formation is highly complex as it adds physical and chemical processes to the complex biological processes involved [15].

With this modern analogue study several goals are pursued for Lake Montcortès: (i) to identify the main limnological and sedimentological mechanisms leading to varves formation; (ii) to establish in which season(s) the deposition of individual calcite sub-laminae occurs and (iii) to determine if the obtained results are coherent with the varve structure and previously proposed mechanisms of their formation [26].

Elevated Ω values were recorded in the epi- and metalimnion Lake Montcortès from late spring to early fall, suggesting that calcite precipitation mostly occurred during this period. The deposition rates were highest in summer when photosynthetic organisms were most active (Figure 2A,B). Additionally, higher POC values were observed during this period, indicating that the main origin of calcite crystals were newly formed autochthonous $CaCO_3$ precipitates (Figure 2C).

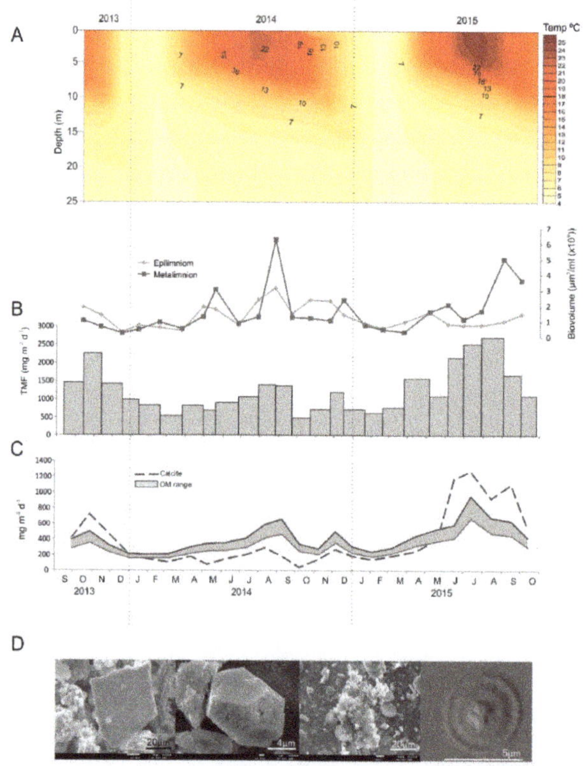

Figure 2. Summary of monthly water column and sediment trap monitoring. (**A**) Water column temperature at 1 m resolution. (**B**) Total mass fluxes (TMF) of monthly sediment traps compared with the phytoplankton biovolume of epi- and metalimnion. (**C**) Calcite fluxes versus organic matter (OM) fluxes of sediment trap material. (**D**) Well-formed blocky and rhombohedral calcite crystals, calcite crystals and diatom frustules and one of the small centric diatoms found in sediment trap material. Redrawn and modified from [18].

In fact, the presence of blocky, polyhedral calcite crystals was confirmed in all sediment trap samples and showed a clear seasonal pattern, being smaller when formed in spring and summer

(Figure 2D). Deposition of smaller crystals was not only concurrent with Ω, but also with larger fluxes of very tiny diatoms (~10 μm), suggesting that phytoplankton cells could be acting as calcite condensation nuclei under enhanced calcite Ω, pH and productivity conditions. Other sources of calcite input, such as detrital carbonates from the catchment, were unimportant for the studied period.

Other seasonal events are precipitation of calcite and OM (Figure 2C). Both variables displayed maximum fluxes in summer to fall. However, calcite precipitation shows the lowest values from winter to spring and was exceeded by OM deposition suggesting a match between the timing of contemporary precipitation of calcite and OM deposition and the alternating couplets of white calcite and brownish organic sublayers of the sediment which would correspond to summer/fall and winter/spring, respectively. These results did not completely back previous paleolimnological inferences that assumed that the light calcite layer was formed in spring/summer and the organic layer was formed in fall/winter [26]. Because of the lack of local or regional modern analogue studies, the authors relied on available literature [17] to explain the observed mechanisms of varve formation.

According to our results, the process of varve forming during the period of study would probably match one of the varve patterns described in the sediment record of Lake Montcortès, once deposited although this remains to be further confirmed. The seasonal calcite-crystal distribution observed in our quarterly sediment traps would produce calcite sub-layering with coarsening upward, i.e., fine-grained calcite crystals deposited in summer and coarse-grained calcite crystals in fall. This texture is often observed in the sedimentary sequence from CE 1350 to 1850 [26]. Modern analogue studies of crystal sizes in calcite sublayers may allow inferring environmental conditions at seasonal resolution, which would be of paleoclimatic significance.

4.2. Is Contemporary Seasonal Deposition of Pollen Consistent with a Two-Layer Varve Model?

The formation of annually laminated lake sediments relies on seasonal deposition linked to climate and on the related flux of biotic and abiotic matter of multiple autochthonous and allochthonous sources to the sediment. Thus, pollen from the catchment and beyond arrives and is buried in lacustrine sediments. Pollen data have been widely used to reconstruct past climate change and have provided insights into the influence of intra-annual weather conditions to pollination patterns. We now argue that combining pollen data from sediment traps and varved lake sediments with modern meteorological data may produce long, high-resolution ecological time series. An example of what can be achieved with this approach are the continuous, long-term and high-resolution climate series obtained by associating paleoclimate data derived from tree rings and other similar proxies with instrumental climate measurements, at annual and seasonal resolutions [39]. To empirically support this hypothesis, this study is primarily intended to identify modern seasonal depositional patterns of pollen in sediment traps deployed in the varved Lake Montcortès that would be useful for interpreting subfossil pollen records from the same lake.

Pollen maxima were recorded during spring, concurrently with the flowering season of the involved taxa, and preceded the temperature and precipitation maxima in summer by one month (Figure 3). Total pollen influx and taxonomic composition show a strong seasonal signal for the study period, peaking positively during spring/summer and negatively during fall/winter. The major components of the pollen assemblages, *Pinus* (pine) and *Quercus* (oak), dominate the pollen counts and match seasonal trends in temperature and precipitation, with *Pinus* almost tripling *Quercus* percentages in spring/summer but falling to similar levels in fall/winter. As far as the remaining species are concerned, most significant differences were observed in *Plantago, Chenopodium, Typha/Sparganium*, Cyperaceae, *Fraxinus* and *Juniperus/Cupressus*, which prevailed in the spring/summer assemblage, and *Cannabis* and *Corylus*, which were more abundant in the fall/winter assemblages (Figure 3). Pollen of *Cannabis* (hemp) prevailed during the fall. *Cannabis* is a cultivated plant with pollen being present and fairly abundant in the surroundings of Montcortès for the last 1200 years. Nonetheless, we still have not been able to locate the exact source of that pollen [33,34].

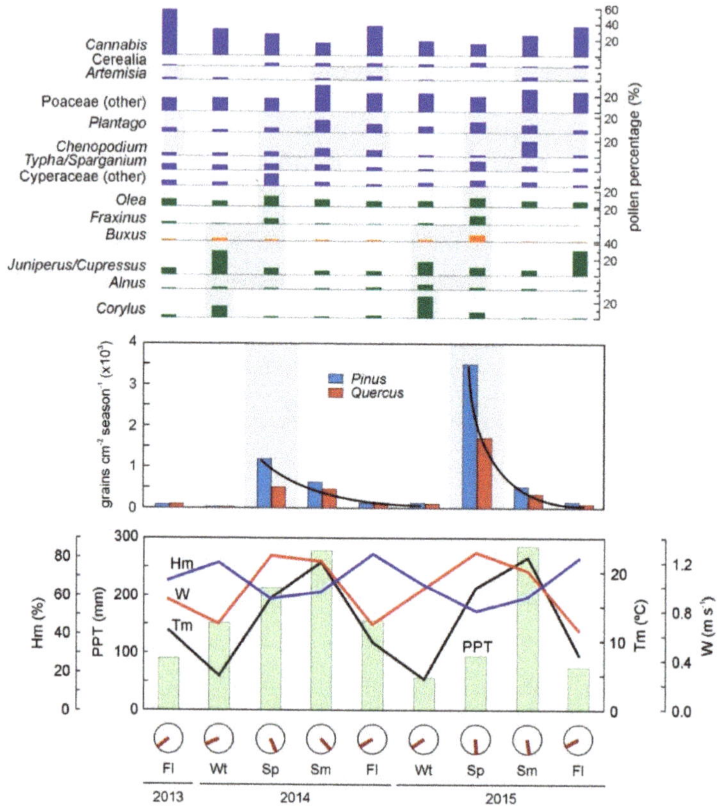

Figure 3. Summary of palynological results showing the relationship among the most relevant meteorological parameters (lower panel), the pollen influx of the superabundant pollen types (middle panel) and the percentages of selected pollen types (upper panel). Average temperature (Tm), relative humidity (Hm) and wind velocity (W) are represented by lines, total precipitation (PTT) is represented by bars. The predominant directions of the wind (Wd) are shown as circles (Sp: spring, Sm: summer, Fl: fall, Wt: winter). Pollen percentages were calculated excluding the abundant *Pinus* and *Quercus*. Taxa are ordered by their respective flowering seasons (grey bands) from bottom to top and from left to right. The flowering season of all species of the different genera present in the Montcortès region [40] has been considered. Trees are shown in green, shrubs are in orange and herbs are in blue. Cultivated plants, such as Cerealia, *Cannabis* and families including many genera (Poaceae, Cyperaceae), are located based on their pollen patterns due to the difficulty of establishing a definite flowering season. Redrawn and modified form [19].

Canonical correspondence analysis (CCA) shows the first two axes accounting for 70.7% of total variance. The strongest gradient coincides with axis 1 (56.8% of the total variance), which was highly and positively correlated with relative humidity and atmospheric pressure and negatively correlated with wind velocity (Figure 4). Along this gradient, pollen samples became aligned from spring (left) to winter (right) suggesting a seasonal succession with summer and fall occupying intermediate positions. Pollen taxa were ordered according to the same gradient. The spring/summer group was highly correlated with temperature, precipitation and wind velocity. The fall/winter group was also highly correlated with wind direction from WSW.

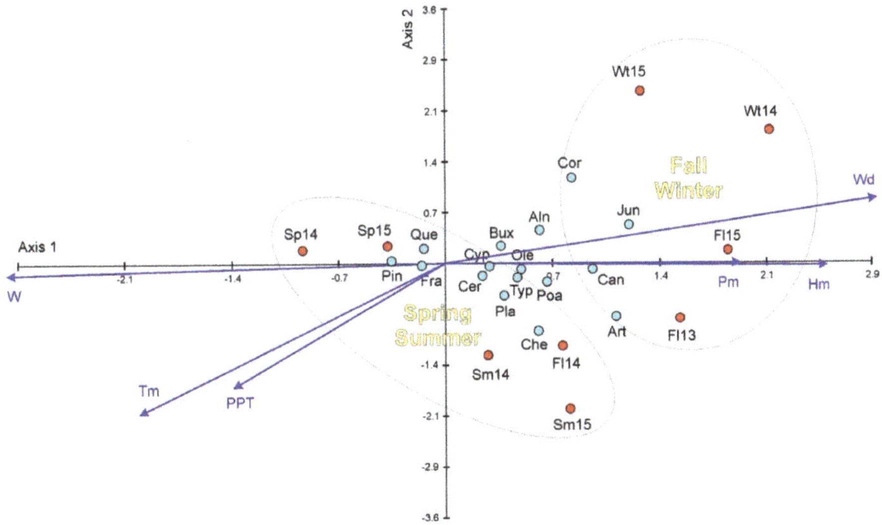

Figure 4. Canonical correspondence analysis (CCA) biplot using the scores of the first two axes accounting for 70.74% of the total variance. Samples are represented by red dots and pollen taxa (for abbreviations see Figure 3) are represented by blue dots. Samples are identified seasonally as follows: Sp—Spring, Sm—Summer, Fl—Fall, Wt—Winter. Meteorological parameters are abbreviated as in Figure 3: Tm—average temperature, Hm—average relative humidity, Pm—average pressure, W—average wind velocity, PPT—average precipitation, Wd—average wind direction. Pollen types: Aln—*Alnus*, Art—*Artemisia*, Bux—*Buxus*, Can—*Cannabis*, Cer—Cerealia, Che—*Chenopodium*, Cor—*Corylus*, Cyp—Cyperaceae (other), Fra—*Fraxinus*, Jun—*Juniperus/Cupressus*, Ole—*Olea*, Poa—Poaceae (other), Pin—*Pinus*, Pla—*Plantago*, Que—*Quercus*, Typ—*Typha*. Redrawn and modified form [19].

Specific aspects of pollen sedimentation require further attention. For example, a lag in pollen sedimentation was observed between production and deposition throughout the year. This lag can have several causes; for example, the influence of internal water dynamics and resuspension, or the fact that soils can be washed into the lake along several successive months after pollen deposition. With respect to lake internal processes, it is worth mentioning that strong thermal water column stratification occurred from March to November 2014 coinciding with the main pollen production period. This stable stratification could have slowed pollen sedimentation rates, whereas in winter vertical mixing of the water disrupted the thermal stability in January and February 2015, facilitating a release of pollen to the lake bottom several months after their production.

The similarity of sedimentation patterns between *Pinus* and *Quercus* pollen was unexpected, because their respective pollen grains show important morphological differences that make their capabilities for air suspension significantly distinct. However, once the pollen is submerged in the lake water of Lake Montcortès, the settling of both types of pollen was quite similar even during summer, when the thermal stratification was stable. This finding suggests that internal lake dynamics neutralise the effect of such differences and that resuspension or catchment runoff may become more important about pollen sedimentation. In general, the seasonal pattern in pollen deposition coincided well with the suggested two-layered varve model. In summary, pollen analysis was able to identify two well differentiated modern assemblages, one corresponding to spring/summer and the other representing fall/winter. This matches well to the seasonal pattern identified in sedimentological (varve) studies [41]. Thus, utilising appropriate transfer functions supports the hypothesis that the combination of pollen data from sediment traps, varved lake sediments and modern meteorological data may produce long and high-resolution ecological time series.

4.3. Revisiting Meromictic Lake Montcortès: Has the Mixing Regime Changed?

Oxygenation of lake water relies on oxygen diffusion from the atmosphere, addition from autochthonous primary production and external inputs of oxygenated water. Lake mixing is the main mechanism, which distributes oxygen throughout the water column and is therefore central for lake health. Mixing regimes can shift with climate changes [42] and human activities [43]. Recently, dissolved oxygen depletion leading to hypoxia or even anoxia has become a matter of concern around the world, and there is a high potential for deterioration under the current global warming [44]. Furthermore, it is suggested that climate change will favour cyanobacteria, which form unpleasant blooms and produce toxins that are harmful to humans [45].

At the same time, obtaining time series of instrumental oxygen records that extend beyond a century is problematic, thus, reducing the possibility of recording long-term changes in oxygen variations that could be associated with climate or human influence. One seminal exception is Lake Zürich, where 50 year long-series of temperature and oxygen profiles corroborated that the increase in the thermal stratification period tends to favour reduction in the homeothermal brake and consequently favours the onset of hypoxia or anoxia [46,47]. Where concurrent series of oxygen and temperature are not available, the combination of proxies of oxic/anoxic conditions with high-resolution paleolimnological records can result in a useful surrogate to assess long-term changes in oxic/anoxic shifts through time.

The varved sediment of LM appears to be highly suitable to provide such high-resolution data. Since redox processes depend on the biogeochemical conditions of lakes and the chemical characteristics of the involved compounds, we firstly conducted a modern analogue study consisting of a monthly survey of main redox indicators, i.e., DO, Fe, Mn and marker pigments [48–51], between October 2013 and October 2016. The same type of marker pigments was also examined in sediments: isorenieratene (Chlorobiaceae) as a tracer of euxinia, okenone (Chromatiaceae) as indicators of anoxia and oscillaxanthin (Oscillatoriales) as a general marker of Cyanobacteria that often points to increased eutrophic conditions. Furthermore, we used selected elemental ratios of the sedimentary record to survey shifts in the oxic/anoxic conditions of the lake over the last 500 years: bromine (Br) as a tracer of organic matter [52], the Ca/Ti and Si/Ti ratios as indicators of biologically mediated calcite and silica production, respectively, and the S/Ti and S/Fe ratios as indicators of the presence of sulphur compounds [53].

Lake Montcortès is reported to be meromictic for years [54]; nonetheless, we observed brief episodes of complete mixing during the winters of 2013 and 2014 that triggered precipitation of Fe and Mn oxides. With the onset of anoxic conditions, precipitated Mn and Fe oxides started to redissolve and sulfur bacteria grew in the hypolimnion under anoxic conditions, when Fe and Mn reached maximum concentrations. The oscillaxanthin was absent in the water column during the sampling period. Such conditions endured over the year 2016, which was a non-mixing year. These results confirmed that redox proxies were working well and provided the information required to properly interpret the sedimentary record. Applying this knowledge to the results obtained from sediment analyses, we were able to distinguish four prevailing situations on the basis of selected redox proxies: (A) years with abrupt and substantial sediment input (turbidites); (B) years with mixing and oxygenation of the water column; (C) years with strong stratification, anoxia, intense activity of sulfur bacteria and increased biomass production; (D) years showing stratification and anoxia, but relatively low biomass production (Figure 5). Interestingly, approximately 45.3% of the years were monomictic years, which mostly occurred in groups of several consecutive years. Meromictic years show a similar pattern. Most (A) years coincide with the climatic instability of CE 1850–1899. The (B) years were rather scattered but were best represented between CE 1820–1849. Most (D) years happened from CE 1500 to 1820, when human activities were locally most intense for the studied period. Almost all (C) years belong to the 20th century. More than 90% of the years with climatic instrumental records belonged to B and C.

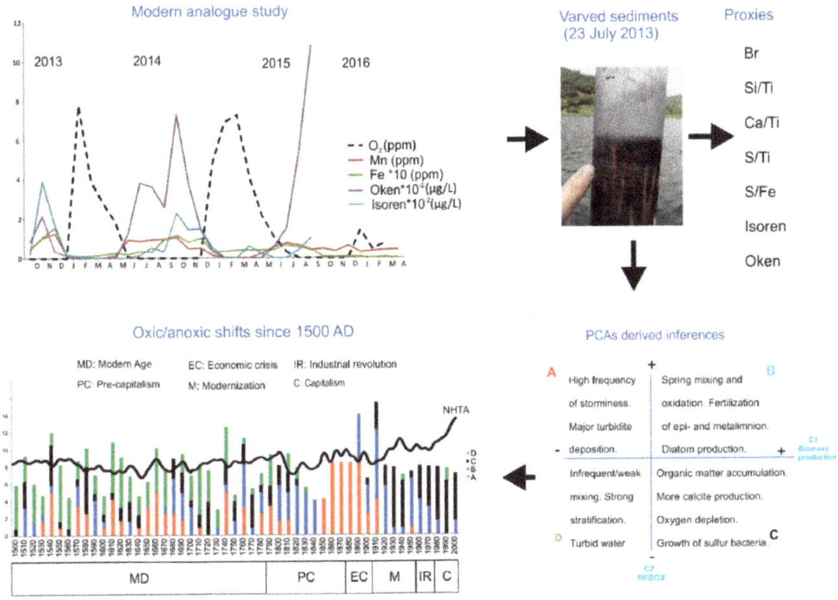

Figure 5. Graphical summary of the modern analogue study about oxygen shifts and inferred results: current seasonal variation of mean values of Fe, Mn, O_2 and phototrophic bacteria pigments okenone (oken) and isorenieratene (isoren) in hypolimnetic waters of Lake Montcortès; varved sediments and proxies obtained; graphic synthesis of inferences derived from Principal Component Analyses; decadal evolution of conditions represented by the subsets A, B, C and D, since CE 1500; bars show the percentage of years that fall within a particular decade. NHTA: Northern Hemisphere temperature anomaly [39]. Redrawn and modified form [20].

4.4. Are GDGTs Promising Indicators of Seasonal Temperature Shifts?

Branched glycerol dialkyl tetraethers (brGDGTs) are bacteria derived and widespread lipids of terrestrial and aquatic environments. The global distribution of brGDGTs in soils and peats has been associated with past temperature, and they are therefore used as proxies of mean annual temperature (MAT proxies) in such environments [55]. However, in some regional studies, brGDGTs indices appeared to be influenced by other variables, e.g., precipitation, humidity and soil properties [56]. This proxy was originally calibrated against annual averages of environmental variables. However, a hypothesis arose that these proxy estimates are biased towards particular seasons, because the bioproduction of brGDGTs is enhanced under the most favourable conditions, which occur mostly in summer.

To answer whether soil brGDGT proxy estimates are seasonally biased, the brGDGT distributions and the brGDGT-derived MAT estimates were examined in settling particulate matter (monthly traps) and surface soil samples from the catchment area of Lake Montcortès (MAT: −3.3 to 17.6 °C). No clear-cut seasonal pattern of the brGDGT distribution has been found in soils (Figure 6a), probably because of the slow turnover time in terrestrial environments, which is on timescales of decades or longer. These results confirm previous findings from mid-latitude soils showing that brGDGTs' distribution and some of the brGDGT-derived proxy measurements are relatively stable through the year [57]. However, we found that annual shifts in abundance of brGDGTs were controlled by the variability of specific regional factors, i.e., soil humidity and their pH [13].

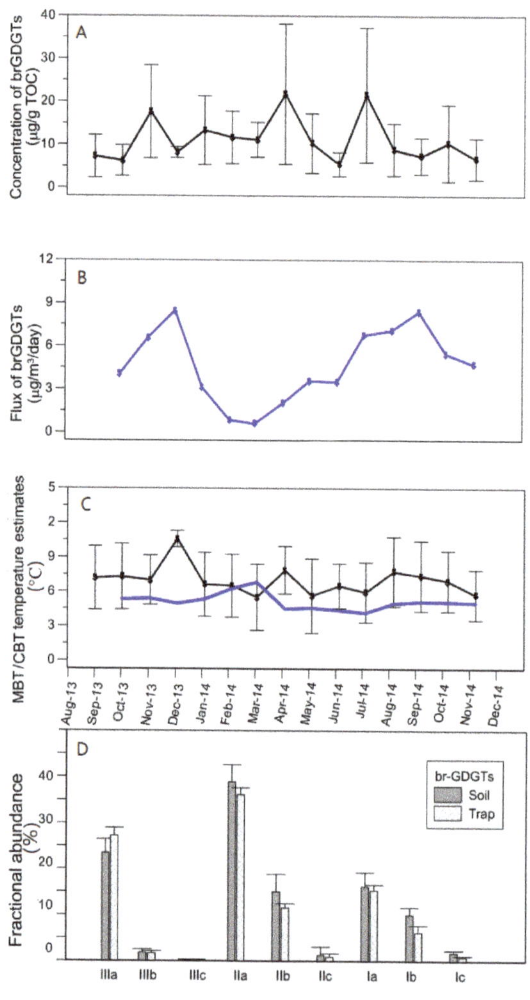

Figure 6. (**A**) Concentration of soil brGDGTs, (**B**) brGDGT flux of particles in the sediment trap at 20 m water depth, (**C**) brGDGT-based temperature estimates using the MBT/CBT (methylation/cyclisation) proxy in soils (black line, average from three soils in the catchment around the lake) and particles from the sediment trap (blue line). (**D**) Fractional abundance of brGDGT distribution in soils and sediment traps. Redrawn and modified from [21].

The results record seasonal trends in sediment flux from the catchment area. In the case of particulate matter, heavy rain is the main factor influencing the brGDGT abundance, resulting in seasonal variations of brGDGT flux (Figures 3 and 6b).

In terms of modern analogues, this study offers evidence that the signatures of brGDGTs in depositional environments are representative of soils in the catchment, and that the accuracy of the derived temperature estimates based on methylation/cyclisation (MBT/CBT) proxies in soils depends on soil properties, that in turn depend on regional factors. Additionally, those brGDGT signatures of soils in the catchment probably represent average environmental conditions over decades or longer, and this observation means that any derived proxy reconstructions can be used only to infer variability in environmental variables over the same timescales. The absence of seasonality in brGDGT proxies from the sediment traps Lake Montcortès is due to the lack of significant "in situ" lacustrine production of brGDGTs (Figure 6c). The most straightforward explanation is that the main sources of brGDGTs are catchment soils, and that their non-seasonal signals are transferred to the settling particles, which have a distribution of brGDGTs that appear to be a weighted mean of soil signals (Figure 6c). In addition, similar patterns of brGDGT fractional abundance in soils and sediment traps confirm soil-related sources to the particles in sediment traps (Figure 6d). It remains unclear if downcore variability in brGDGTs at an annual scale will also capture changes derived from in situ production, and if derived brGDGT proxies will be suitable to build high-resolution MAT reconstructions.

4.5. From Contemporary Phytoplankton to Subfossil Pigments, What Can We Learn about Community Changes?

In general, qualitative and quantitative relationships between modern climatic variables, aquatic primary producers and their pigments and subfossil marker pigments of the sediment have received little attention despite their potential to build reliable climatic proxies. To help fill this gap, we performed a modern analogue study to determine how phytoplankton and marker pigment information compare with the sedimentary record. For this purpose, we applied an analogue matching (AM) technique, which is concerned about identifying contemporary sites that most closely match the species assemblage identified in the past [7].

We assessed and compared the annual cycle of phytoplankton in the epi- and metalimnion and of marker pigments in the epi-, meta- and hypolimnion, and also in a sediment trap deployed at 20 m water depth. Redundancy Analysis (RDA) was applied to identify potential associations between the phytoplankton taxa identified by microscopy. Then, we performed a similarity analysis in order to examine expected bias between modern and subfossil samples (CE 1493–2013) in terms of presence or absence of common pigment markers.

The total biovolume of phytoplankton depicted a regular annual cycle with growing peaks in early spring and summer and low biomass in winter (Figure 2B). Phytoplankton biovolume of the metalimnion was highest in summer ($>6 \times 10^6$ μm^3/mL). Centric Bacillariophyceae and Chlorophytes follow one another as the dominant taxa in 2013–2014, whereas smaller centric Bacillariophyceae took over during 2014–2015 growing periods (Figure 7).

The RDA analysis allowed to identify marker pigments more tightly associated with each taxonomic group (Figure 8). Chlorophyta, Dinophyta, Cyanoprokaryota (Cyanobacteria and Cyanophyta), centric Bacillariophyceae, Volvocales and Cryptophyta biovolumes explain a significant percentage of variance in marker pigment concentration, while the percentage variance explained by the other taxonomic groups was insignificant. Each taxonomic group was associated with specific marker pigments. The biovolume of Chlorophyta was associated with the marker pigments neoxanthin, lutein, chlorophyll-b and zeaxanthin, the biovolume of Dinophyta mainly with peridinin, diadinoxanthin, fucoxanthin, Chl-c1 and c2, although Chl-c2 was also partially associated with the biovolume of Cryptophyta. The biovolume of Cryptophyta was mainly associated with alloxanthin and a-carotene, both partially explained by pennate Bacillariophyceae. We know from the literature that some of the pigments associated with Dinophyta are also present in Bacillariophyceae (e.g., fucoxanthin, diadinoxanthin, Chl-c1 and Chl-c2), which is coherent with the observed RDA biplot association. Cyanobacteria biovolume was associated with myxoxanthophyll-like, while association with marker pigments of phototrophic bacteria is due to co-occurrence in the samples with these groups. These data are essential to identify pigments with a higher potential to be used as modern analogues in the sediment record.

All this information is relevant to better identify pigments from sediment traps and use them as modern analogues for the sediment record following taxonomic affinity criteria and preservation.

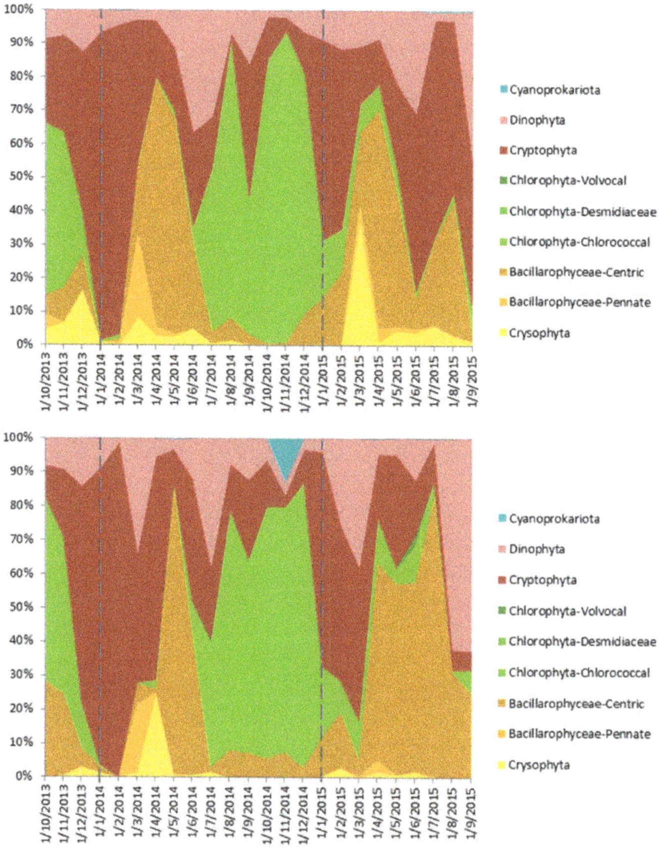

Figure 7. Taxonomic phytoplankton groups with their biovolume and relative dominance (%) in the epi- (**upper graph**) and the metalimnion (**lower graph**) from October 2013 to September 2015 (abscissa).

The concentration of most representative marker pigments was calculated for the entire water sampling period (2013–2015), in order to compare their spatial variations between epi-, meta- and hypolimnion, as well as with their deposition in the sediment trap (Figure S1, Supplementary Materials). Marker pigments of Bacillariophyceae (fucoxanthin: Figure S1c) and Dinophyta (diadinoxanthin: Figure S1d) were produced and deposited in higher amounts in spring and summer while Chlorophyta marker pigments (Chl-b, lutein and zeaxanthin: Figure S1e,f,h) were produced mainly in fall, and marker pigments of Cryptophyta (alloxanthin: Figure S1g) were produced throughout the year with the exception of fall. Zeaxanthin is a marker pigment also indicative of Cyanobacteria and decoupled from parent marker pigments of Chlorophyta during summer, it indicates a higher contribution of Cyanobacteria during summer months (Figure S1). Marker pigments of photosynthetic sulphur bacteria were only found in the hypolimnion during summer and fall (Figure S1i,j).Most marker pigments from the water column were also identified in the sediment record. A list of the marker pigments co-occurring in the water column and sediment record is given in Table S1 (Supplementary Materials). Their taxonomic affinities are also shown and inform about their likely biological sources, which are consistent with the taxonomic phytoplankton groups of the epi- and metalimnion (Figure 7). To gauge the similarity between water column and sediment samples we applied the Sorensen index (S) on a similarity matrix built with binary data. This very simple similarity index provides a greater "weight" to species common to the samples than to those found in only one sample. According to S, epi-, meta- and hypolimnetic samples resemble at ≥80%, whereas meta and hypolimnetic samples resemble each other relatively more (92%), likely because of lower oxidation rates during anoxic conditions of the hypolimnion. Interestingly, when compared with sediment samples of the top of the core (CE 2007–2013 E), S diminished abruptly to 0.5, that is, the bias was 50% (Figure 9). This decrease was followed by a slower decrease of S, until values of 30% (circa 1850) and its posterior stabilisation at values of 40–60% that lasted circa two centuries. A Venn diagram highlights that past communities only have a partial modern analogue (Figure 9) [5]. These results indicate that approximately 50% of the marker pigments in the water column were destroyed between deposition and permanent burying in the first centimetres of the sediment record, whereas the remaining 40–60% were accurately represented over the period of study. This pattern is roughly consistent with the three phases of sediment loss proposed by [48]: (1) rapid oxidation, enzymatic metabolism and digestion by herbivores through the water column (half-life T1/2 = days) would account for a breakdown of pigments while they sink, (2) slower post-depositional loss in surface sediments would result from structural rearrangements, release of labile compounds and further oxidation (T1/2 = years) and (3) once buried pigment degradation would have continued at a very slow pace (T1/2 = centuries).

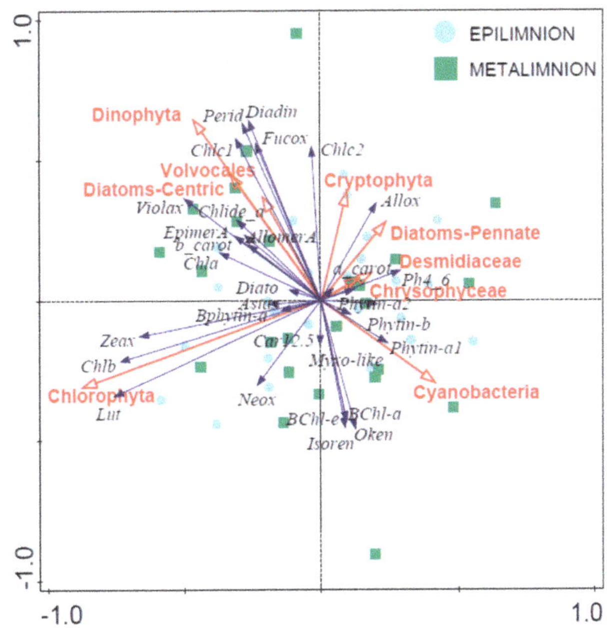

Figure 8. Redundancy analysis (RDA) between phytoplankton taxa (biovolume; $mm^3 \cdot L^{-1}$) and marker pigments (concentration; $\mu g \cdot L^{-1}$). Red arrows are explanatory variables (53% of the variance in pigment concentration was explained by phytoplankton taxa), blue arrows are response variables (pigments). Diadin: diadinoxanthin; Fucox: fucoxanthin; Chlc2: chlorophyll-c2; Allox: alloxanthin; a-carot: a-carotene; Ph4_6: unknown phorbin; Phytin-a1,a2: phaeophythin-a1, a2; Phytin-b: phaeophythin-b; Car12.5: unknown carotenoid; Myxo-like: myxoxanthophyll-like; BChl-a: bacteriochlorophyll-a; BChl-e: bacteriochlorophyll-e; Oken: okenone; Isoren: isorenieratene; Neox: neoxanthin; Lut: lutein; Chlb: chlorophyll-b; Zeax: zeaxanthin; Bphytin-a: bacteriophaeophythin-a; Astax: astaxanthin; Diato: diatoxanthin; Chla: chlorophyll-a; b-carot: b-carotene; EpimerA: Chl-a epimer; AllomerA: Chl-a allomer; Violax: violaxanthin; Chlide_a: chlorophyllide-a; Chlc1: chlorophyll-c1; Perid: peridinin.

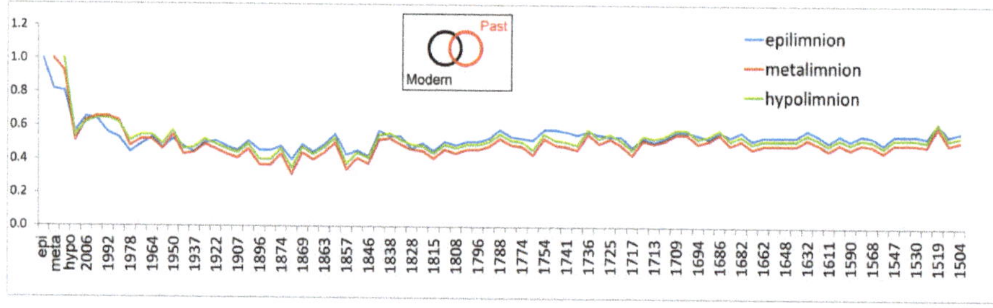

Figure 9. Comparison of Sorensen's similarity index (S) between modern and sediment samples. A Venn diagram comparing modern and past pigment assemblages is shown in the upper right corner. Epi-, meta-, hypo- refer to current in-depth layers of the water column. Abcissa: dates inform about the age of the sediment samples, Ordinate: S values.

Finally, Table S2 (Supplementary Materials) shows which marker pigments were found to be exclusive from present-day samples (6) and which were found only in subfossil samples (7). Some of the unspecific chlorophyll derivatives are produced under distinct environmental conditions and by different mechanisms with time, chiefly through losses of Mg^{+2}, herbivory, viral attack and enzymatic catalysis [48]. Noteworthy, however, is the presence of four Cyanobacterial markers in the subfossil record that were not present in modern samples, suggesting either a detection limit for the identification of these pigments in the water column or that past lake conditions were favourable to a high diversity of Cyanobacterial groups including N_2-fixing. In fact, massive blooms of filamentous *Planktothrix rubescens* De Candolle ex Gomont have been reported to have flourished in Lake Montcortès in the 1970s [54]. However, based on microscopical counts and marker pigment detection, Cyanobacteria seem to constitute only a low percentage of the phytoplankton currently thriving at Lake Montcortès (Figure 6), an issue that needs further confirmation.

5. Discussion, Conclusions and Guidelines for Future Research

To date, most paleoenvironmental reconstructions for Lake Montcortès have been performed using conventional methods and models, which were originally thought for lower resolution studies and higher latitudes. They are therefore constrained to the supra-decadal level. However, it is clear that the sublayers of varves host invaluable, high resolution information that still has to been retrieved. Our modern analogue studies provide detailed information about fine-tuned processes that take place at annual and sub-annual scales and show how a suite of widely used proxies respond at the same temporal resolution. We expect that each proxy is physically deposited in a varve sublayer in temporal correspondence with the processes it represents. Being able to put this exact match in evidence is expected to boost the interpretative power of future high-resolution paleoenvironmental reconstructions.

The main contribution of this review is to show the advantages and limitations of a multiproxy modern-analogue approach in Lake Montcortès as a case study. This has not been published previously in any other papers. There exist other studies with a similar approach, e.g., lake Zabińiski (Poland) [58] or Lake La Cruz (Spain) [59–62], but results in every case are different because sedimentary and biogeochemical processes differ among lakes. Obviously, results of one particular lake may not be directly applicable at other lakes. Our results demonstrate that current lacustrine processes of Lake Montcortès involving calcite precipitation, phytoplankton and bacterial growth, oxygenation of the water column and pollen production, have a clear seasonality and also a year-to-year variability, depending on subtle interannual climatic variations. Therefore, it is reasonable to expect finding the corresponding fingerprints of these processes at the varve (sublayer) levels. Interestingly, the brGDGT production in soils is not seasonal, while the particle flux into the lake is. The similarity of brGDGT distributions indicated the same sources of brGDGTs in soils and lakes. The brGDGT-estimated temperature reflected the interannual average conditions from the soil records. In order to provide an integrated overview of the basic functioning of Lake Montcortès related to our modern analogue studies, a synthesis of the main processes and their spatial and temporal course is presented (Figure 10). This figure is intended to serve as a guide for high resolution paleoecological and paleoenvironmental reconstructions and derived inferences in lake Montcortès and other similar lakes. It shows how limnological variables and processes give raise to the proxies that will be used in these reconstructions, the transit of the latter from present to past conditions, and where in the varve structure would these proxies be found, according to the season where they are produced and settled down (time-lag = 0 is assumed). In this figure, the limnological variables under study are listed (column 1); moreover, the contemporary processes (column 4) that occur in the lake or lake catchment over the year (columns 2, 3). Columns 5 to 8 show the seasons when these processes occur (blue) and when they are more intense (dark blue). It is also indicated which varve sublayer is representative of each season. White and a dark sublayer correspond to summer and winter, respectively, dark-white transitions to spring and white-dark transitions to fall. Column 9 displays degradation and diagenetic mechanism that are expected to act upon the limnological variables inside the sediment and column 10 the derived proxies. Column 11 suggests which varve sublayers should be targeted to retrieve these proxies for high resolution reconstructions, accepting that one single varve contains all seasonal information needed, i.e., in correspondence with white and dark sublayers and their transitions. We also assume that highly precise sampling methods are available in every case. This is an especially sensitive issue in the case of Lake Montcortès where varves are relatively thin, e.g., total mean varve thickness averages 1.3 mm between the 14th and 20th century, with calcite sublayers 0.35 mm and organic sublayers 0.34 mm [26], and probably in other lakes as well. Note for example that by now, neither diatoms nor pollen or pigments could be extracted from a single varve and even less from a sublayer, due to methodological limitations.

Modern-analogue variables	Catchment	Lake	Current process	Season				Transformation in the sediment (diagenesis)	Sedimentary proxy	Expected main target sublayers
				Sp (tr d-w)	Sm (w)	Fl (tr w-d)	Wt (d)			
Calcite saturation index	x	x	Endogenic Ca precipitation, primary production.					Calcite deposition and accumulation.	Ca/Ti, calcite laminations.	w; tr w-d
Large calcite crystals		x	Authigenic formation of calcite crystals, low T°C.					Accumulation and growth.	Larger calcite crystals.	tr w-d; d
Small calcite crystals			Biogenic formation of calcite crystals, high T°C.					Accumulation and growth.	Smaller calcite crystals.	tr d-w; w
Particulate organic C (POC)	x	x	Lacustrine primary production.					Organic carbon accumulation/oxydation.	Br, S/Ti	w; tr w-d
Phytoplankton biovolume		x	Phytoplankton growth and succession.					Degradation.	Subfossil pigments.	w; tr w-d
Diatom species		x	Diatom blooms.					Accumulation of frustules.	Subfossil diatoms.	tr d-w
Oxygen		x	O_2 consumption and supply .					O_2 consumption.	Oxidised/reduced mineral form	tr d-w; tr w-d
O2, Mn,Fe	x	x	Hypolimnetic oxidation , mixing.					Precipitates of Mn, Fe oxides/hydroxides.	Fe/Mn, Fe/Ti, absence of anaerobic pigments.	d
O2, Mn,Fe	x	x	Hypolimnetic reduction, anoxia.					Redissolution of Mn, Fe oxides/hydroxides	Fe/Mn, Mn/Ti, presence of anaerobic bacterial pigments.	w; tr w-d
POLLEN										
Pinus	x		Flowering season					Partial degradation	Pollen abundance	tr d-w; w
Plantago	x		Flowering season					Partial degradation	Pollen abundance	w
Chenopodium	x		Flowering season					Partial degradation	Pollen abundance	w
Typha/Sparganium		x	Flowering season					Partial degradation	Pollen abundance	tr d-w; w
Fraxinus	x		Flowering season					Partial degradation	Pollen abundance	tr d-w
Juniperus/Cupressus	x		Flowering season					Partial degradation	Pollen abundance	tr w-d; d
Cannabis	x		Flowering season					Partial degradation	Pollen abundance	tr w-d; d
Corylus	x		Flowering season					Partial degradation	Pollen abundance	d
Total pollen	x		Flowering season					Partial degradation	Pollen abundance	tr d-w; w
LIPIDS										
brGDGT in soils	x		Bioproduction of brGDGT						brGDGT	
brGDGT in traps		x	brGDGT drainage					Accumulation and degradation	brGDGT	tr w-d
PRIMARY PRODUCERS										
Chlorophyta biovolume		x	Pigment production					Degradation and oxidation	Production of neox, lut, chl-b, zeax.	w; tr w-d
Dinophyta biovolume		x	Pigment production					Degradation and oxidation	Production of perid, diadinox, fucox, Chl-c1 and c2 .	tr d-w; w; tr w-d; d
Cryptophyta biovolume		x	Pigment production					Degradation and oxidation	Production of allox, a-carot,	all sublayers
Bacillariophyta		x	Pigment production, silicate consumption					Degradation and oxidation	Production of allox, a-carot, fucox, diadinox, Chlc1, Chlc2. Diatom frustules.	tr d-w
Cyanobacteria		x	Pigment production					Degradation and oxidation	Production of myxo-like.	tr w+d
Phototrophic bacteria		x	Pigment production					Degradation and oxidation	Production of oken, isoren.	w; tr w-d

Figure 10. Synthesis of main modern analogue variables and processes at Lake Montcortès, derived proxies and their expected target sublayers. White (w); dark (d): white-dark transitions (w-d tr) and dark-white transitions (d-w tr) between sublayers. Sp: spring, Sm: summer, Fl: Fall, Wt: winter.

Finally, we provide some guidelines for future research at Lake Montcortès and similar lakes.

i. Calcite: There is evidence that a temperature increase enhances the amount of calcite precipitation and flux to the sediment. This relationship may be an important factor, when considering the role of calcite as a sink of atmospheric CO_2 in the context of global warming and needs further examination. On the other hand, if the size of endogenic calcite crystals varies seasonally, inferences at the sub-annual scale may be derived from calcite crystals contained in sublayers of varves.

ii. These hypotheses encourage the execution of further modern analogue studies. Hence, three improvements are suggested: (1) increase the frequency and duration of sampling, in order to better document shifts in calcite precipitation as related to the current warming trend; (2) install a meteorological station in the lake's catchment, in order to capture local temperature and other meteorological variables that condition water temperature; (3) perform a detailed survey of calcite crystal features retrieved from seasonal traps and from varve sublayers to precisely determine their origin and diagnostic characteristics.

iii. Pollen: Two advantages of pollen studies are (1) the ability of identifying and characterizing seasonal layers even in the absence of varves and (2) the possibility of recording interannual variability and associated meteorological drivers.

iv. Oxygen: Even though our approach cannot provide estimations of past oxygen concentrations, it is a useful tool to connect and compare the dynamics of past and present oxic and anoxic events with annual resolution. However, if past oxygen concentrations are to be inferred, a more specific modern analogue study is required. Perhaps monthly and multiannual monitoring of the evolution of hypolimnetic oxygen concentrations and a concurrently examination of the content of Fe and Mn oxides in surface sediment samples, catches mixing and meromictic years. It should then be possible to establish appropriate transfer functions to estimate dissolved oxygen from most related proxies. If successful, truly long-term oxygen time series linking estimated and instrumental data could be established.

v. Cyanoprokariota assemblages lack a modern analogue, since they are nearly absent today, whereas they were abundant, diverse and pervasive until, at least, the 1970s [54]. This observation is surprising because, to our knowledge, there have been few changes in land use or lake exploitation that could have affected Lake Montcortès' trophic conditions. In contrast, we expect Cyanoprokariota to play a prominent role during the coming warmer years, when Lake Montcortès becomes more intensely stratified. Cyanobacteria marker pigments (myxoxanthophyll and zeaxanthin) have been detected in the water column during monitoring, the lack of correspondence with microscopical counts may be due to the low cell size of species involved, but this will need further studies. Therefore, considering the known deleterious effects of Cyanobacteria on aquatic ecosystems and human health, it would be advisable to continue present-day monitoring. The results would aid in obtaining a long-term picture of cyanobacterial successions and importance and allow deriving future scenarios, in the context of global warming.

vi. brGDGT production in the soils of the lake catchment show no seasonality. However, in sediment trap particles, the brGDGT flux presents a clear seasonality. The seasonal variations of flux rates confirm transport variations from the catchment rather than in situ production of brGDGT in the lacustrine environment. The MBT/CBT temperature estimates indicate that brGDGT signatures in the sediment traps show a mixed signature and originate from surrounding soils. These results suggest that any inferences made from the high-resolution record must be made with caution.

vii. From a methodological point of view, cylindrical traps with seasonal or quarterly recovery yield coherent results in all cases, i.e., total suspended solids, calcite crystals, pollen and spores and particles for brGDGT extractions. However, it is recommended that additional samples of surface sediments at the end of sampling years should be taken, to assess post-depositional transformations of these proxies. In our case, surface sediment samples would have provided complementary information, such as total fluxes to the sediment and subsequent accumulation, interactions at the sediment-water interface and early diagenesis. For example, in the case of calcite precipitation as related to varve formation, initial data of organic matter content may be needed to estimate the degree of sediment compaction with time [63] and for a full understanding of structure and thickness of varve sublayers at any depth. In the end, all this information allows a better connection and comparison between contemporary proxies and their modified version after being buried in the sediment, making the modern analogue technique more powerful.

viii. Sometimes, modern analogue studies performed at a seasonal or annual frequency cannot be fully exploited for the interpretation of sedimentary records, because of a mismatch in resolution between present-day and past samples. This mismatch occurs when a relatively large weight or volume of sediment is needed to extract enough material for proxy analyses, because contiguous varves are too thin and/or must be joined into one sample. In doing so, the resolution of the samples decreases. With time and targeted effort, this disadvantage will hopefully be overcome with the advent of new technologies. In the field of marker pigments, hyperspectral image spectroscopy is progressing quickly and offers a non-destructive and inexpensive approach that permits high resolution. Hyperspectral image spectroscopy is being used in lake sediments, e.g., for concentrations of sedimentary bacteriopheophytin "a", based upon diagnostic spectral properties with high spatial and temporal resolution [64]. A good match of the resolutions of modern and past studies is indispensable to successfully connect both and to better understand the evolution of ecosystems with time.

Supplementary Materials: The Supplementary Materials is available online at http://www.mdpi.com/2571-550X/3/1/1/s1. Figure S1: Comparison of most representative modern marker pigments from the epi-, meta- and hypolimnion as well as their deposition in the sediment trap, on the basis of annual values of the period 2013–2015; Table S1: Marker pigments co-occurring in the water column and sediment record and their taxonomic affinities for lake Montcortès; Table S2: Marker pigments and their taxonomic affinities for Lake Montcortès.

Author Contributions: T.V.-V. conceived and wrote the paper. T.B., V.R., M.d.C.T., M.C., A.R.-M., J.G., J.P.C. and P.L. contributed by refining the text and with data, ideas and figures. E.S., M.d.C.T., N.C., J.G., S.G.-P., P.L., S.G., J.P.C. and T.B. carried out fieldwork and laboratory analyses. J.S. and N.P.-Z. advised on climatic aspects and contributed with meteorological data. All co-authors reviewed the paper. All authors have read and agreed to the published version of the manuscript.

Funding: Research funding was granted by the Spanish Ministry of Economy and Competitivity (MINECO/FEDER) with the projects MONT-500, ref. CGL2012-33665; GLOBALKARST, ref. CGL2009-08145), the Institute of Catalan Studies with the projects PIRIMOD and POLMONT, and the Catalan University and Research Management Agency (AGAUR) with the projects 2014 SGR 1207, 2017 SGR 1116).

Acknowledgments: We acknowledge the Busseing Pallars Company, the Council of Baix Pallars and the cultural association Lo Vent de Port for their direct involvement in the project and their continuous support.

Conflicts of Interest: The authors declare no conflict of interest.

References

1. Gentner, D. Structure-mapping: A theoretical framework for analogy. *Cogn. Sci.* **1983**, *7*, 155–170. [CrossRef]
2. Gentner, D. The mechanisms of analogical learning. In *Similarity and Analogical Reasoning*; Vosniadou, S., Ortony, A., Eds.; Cambridge University Press: London, UK, 1989; pp. 199–224.
3. Hesse, M.B. *Models and Analogies in Science*; University of Notre Dame Press: Notre Dame, IN, USA, 1970; p. 184.
4. Markman, A.B. Constraints on analogical inference. *Cogn. Sci.* **1997**, *21*, 373–418. [CrossRef]
5. Delcourt, H.R.; Delcourt, P. The Paleoecological Perspective. In *Quaternary Ecology: A Paleoecological Perspective*; Chapman Hall: London, UK, 1991; p. 241.
6. Jackson, S.T.; Williams, J.P. Modern Analogs in Quaternary Paleoecology: Here Today, Gone Yesterday, Gone Tomorrow? *Ann. Rev. Earth Planet. Sci.* **2004**, *32*, 495–537. [CrossRef]
7. Simpson, G.L. Analogue Methods in Palaeolimnology. In *Tracking Environmental Change Using Lake Sediments*; Birks, H.J.B., Lotter, A.F., Juggins, S., Smol, J.P., Eds.; Springer Science Business Media: London, UK, 2012; Volume 5, pp. 495–522.

8. Legendre, P.; Birks, H.J.B. From classical to canonical ordination. In *Tracking Environmental Change Using Lake Sediments*; Smol, J.P., Birks, H.J.B., Last, W.M., Eds.; Kluwer: Dordrecht, The Netherlands, 2012; pp. 201–248.
9. Weng, C.; Bush, M.B.; Silman, M.R. An analysis of modern pollen rain on an elevational gradient in southern Peru. *J. Trop. Ecol.* **2004**, *20*, 113–124. [CrossRef]
10. Reuss, N.; Leavitt, P.R.; Hall, R.I.; Bigler, C.; Hammarlund, D. Development and application of sedimentary pigments for assessing effects of climatic and environmental changes on subarctic lakes in northern Sweden. *J. Paleolimnol.* **2010**, *43*, 149–169. [CrossRef]
11. Sachs, H.; Webb, T.; Clark, D.R. Paleoecological transfer funcions. *Ann. Rev. Earth. Planet. Sci.* **1977**, *5*, 159–178. [CrossRef]
12. Massaferro, J.; Larocque-Tobler, I. Using a newly developed chironomid transfer function for reconstructing mean annual air temperature at Lake Potrok Aike, Patagonia, Argentina. *Ecol. Ind.* **2013**, *24*, 201–210. [CrossRef]
13. Pla, S.; Catalan, J. Chrysophyte cysts from lake sediments reveal the submillennial winter/spring climate variability in the northwestern Mediterranean region throughout the Holocene. *Clim. Dyn.* **2005**, *24*, 263–278. [CrossRef]
14. Wittkop, C.; Teranes, J.; Lubenow, B.; Dean, W.E. Carbon-and oxygen-stable isotopic signatures of methanogenesis, temperature, and water column stratification in Holocene siderite varves. *Chem. Geol.* **2014**, *11*, 153–166. [CrossRef]
15. Zolitschka, B.; Francus, P.; Ojala, A.E.; Schimmelmann, A. Varves in lake sediments–A review. *Quat. Sci. Rev.* **2015**, *117*, 1–41. [CrossRef]
16. Rull, V. Time continuum and true long-term ecology: From theory to practice. *Front. Ecol. Evol.* **2014**, *2*, 1–7. [CrossRef]
17. Brauer, A. Annually laminated lake sediments and their paleoclimatic relevance. In *Climate in Historical Time: Towards a Synthesis of Holocene Proxy Data and Climate Models*; Fisher, H., Ed.; Springer: Berlin/Heidelberg, Germany, 2004; pp. 108–128.
18. Trapote, M.C.; Vegas-Vilarrúbia, T.; López, P.; Puche, E.; Gomà, J.; Buchaca, T.; Cañellas-Boltà, N.; Safont, E.; Corella, J.P.; Rull, V. Modern sedimentary analogues and integrated monitoring to understand varve formation in the Mediterranean Lake Montcortès (Central Pyrenees, Spain). *Palaeogeogr. Palaeoclimatol. Palaeoecol.* **2018**, *496*, 292–304. [CrossRef]
19. Rull, V.; Trapote, M.C.; Safont, E.; Cañellas-Bolta, N.; Pérez-Zanón, N.; Sigró, J.; Buchaca, T.; Vegas-Vilarrúbia, T. Seasonal patterns of pollen sedimentation in Lake Montcortès (Central Pyrenees) and potential applications to high-resolution paleoecology: A 2-year pilot study. *J. Paleolimnol.* **2017**, *57*, 95–108. [CrossRef]
20. Vegas-Vilarrúbia, T.; Corella, J.P.; Pérez-Zanón, N.; Buchaca, T.; Trapote, M.C.; López, P.; Sigró, J.; Rull, V. Historical shifts in oxygenation regime as recorded in the laminated sediments of lake Montcortès (Central Pyrenees) support hypoxia as a continental-scale phenomenon. *Sci. Tot. Environ.* **2018**, *612*, 1577–1592. [CrossRef] [PubMed]
21. Cao, M.; Rueda, G.; Rivas-Ruiz, P.; Trapote, M.C.; Henriksen, M.; Vegas-Vilarrúbia, T.; Rosell-Melé, A. Branched GDGT variability in sediments and soils from catchments with marked temperature seasonality. *Org. Geochem.* **2018**, *122*, 98–114. [CrossRef]
22. Corella, J.P.; Moreno, A.; Morellón, M.; Rull, V.; Giralt, S.; Rico, M.T.; Pérez-Sanz, A.; Valero-Garcés, B.L. Climate and human impact on a meromictic lake during the last 6000 years (Montcortès Lake, Central Pyrenees, Spain). *J. Paleolimnol.* **2011**, *46*, 351–367. [CrossRef]
23. Rosell, J. *Geological Map of Spain and Report. Scale 1:50,000, Tremp Sheet (252)*; Instituto Tecnológico Geográfico de España (IGME): Madrid, Spain, 1994.
24. Mercadé, A.; Vigo, J.; Rull, V.; Vegas-Villarrúbia, T.; Garcés, S.; Lara, A.; Cañellas-Boltà, N. Vegetation and landscape around Lake Montcortès (Catalan pre-Pyrenees) as a tool for palaeoecological studies of lake sediments. *Collect. Bot.* **2013**, *32*, 87–101. [CrossRef]
25. Corella, J.P.; Benito, G.; Wilhelm, B.; Montoya, E.; Rull, V.; Vegas-Vilarrúbia, T.; Valero-Garcés, B.L. A millennium-long perspective of flood-related seasonal sediment yield in Mediterranean watersheds. *Glob. Plan. Sci.* **2019**, *177*, 127–140. [CrossRef]
26. Corella, J.P.; Brauer, A.; Mangili, C.; Rull, V.; Vegas-Vilarrúbia, T.; Morellón, M.; Valero-Garcés, B.L. The 1.5-ka varved record of Lake Montcortès (Southern Pyrenees, NE Spain). *Quat. Res.* **2012**, *78*, 323–332. [CrossRef]

27. Corella, J.P.; Benito, G.; Rodríguez-Lloveras, X.; Brauer, A.; Valero-Garcés, B.L. Annually resolved lake record of extreme hydro-meteorological events since AD 1347 in NE Iberian Peninsula. *Quat. Sci. Rev.* **2014**, *93*, 77–90. [CrossRef]
28. Pierrot, D.; Lewis, E.; Wallace, D.W.R. *MS Excel Program Developed for CO2 System Calculations, ORNL/CDIAC-105*; Carbon Dioxide Information Analysis Center, Oak Ridge National Laboratory, US Department of Energy: Oak Ridge, TN, USA, 2006.
29. Battarbee, R.W.; Jones, V.; Flower, R.J.; Cameron, N.G.; Bennion, H.; Carvalho, L.; Juggins, S. Diatoms. In *Tracking Environmental Change Using Lake Sediments*; Smol, J.P., Birks, H.J.B., Last, W.M., Eds.; Kluwer: Dordrecht, The Netherlands, 2001; Volume 3, pp. 155–202.
30. Krammer, K.; Lange-Bertalot, H. Bacillariophyceae. In *Süßwasserflora von Mitteleuropa, Bänder 2,3,4/5*; Gustav FischerVerlag: Stuttgart, Germany, 2000.
31. Lange-Bertalot, H. *Diatoms of the Europe Inland Waters and Comparable Habitats*; Gantner: Ruggell, Germany, 2001.
32. Bennett, K.D.; Willis, K.J. Pollen. In *Tracking Environmental Change Using Lake Sediments*; Smol, J.P., Birks, H.J.B., Last, W.M., Eds.; Kluwer: Dordrecht, The Netherlands, 2002; pp. 5–30.
33. Rull, V.; Vegas-Vilarrúbia, T. Preliminary report on a mid-19th century Cannabis pollen peak in NE Spain: Historical context and potential chronological significance. *Holocene* **2014**, *24*, 1378–1383. [CrossRef]
34. Rull, V.; Vegas-Vilarrúbia, T. Crops and weeds from the Estany de Montcortès catchment, Central Pyrenees, during the last millennium: A comparison of palynological and historical records. *Veg. Hist. Archaeobot.* **2015**, *24*, 699–710. [CrossRef]
35. Rull, V.; González-Sampériz, P.; Corella, J.P.; Morellón, M.; Giralt, S. Vegetation changes in the southern Pyrenean flank during the last millennium inrelation to climate and human activities: The Montcortès lacustrine record. *J. Paleolimnol.* **2011**, *46*, 387–404. [CrossRef]
36. Buchaca, T.; Catalan, J. Factors influencing the variability of pigments in the surface sediments of mountain lakes. *Fresh. Biol.* **2007**, *52*, 1365–1379. [CrossRef]
37. Buchaca, T. *Pigments Indicadors: Estudi del Senyal en Estanys dels Pirineus i de la Seva Aplicació en Paleolimnologia*; Institut d'Estudis Catalans: Catalans, Spain, 2009.
38. Leemann, A.; Niessen, F. Varve formation and climatic record in an Alpine proglacial lake: Calibrating annually-laminated sediments against hydrological and meteorological data. *Holocene* **1994**, *4*, 1–8. [CrossRef]
39. Mann, M.E.; Bradley, R.S.; Hughes, M.H. Northern Hemisphere Temperatures during the Past Millennium' Inferences, Uncertainties, and Limitations. *Geophs. Res. Lett.* **1999**, *26*, 759–762. [CrossRef]
40. Bolòs, O.; Vigo, J.; Masalles, R.M.; Ninot, J.M. *Flora Manual dels Països Catalans*; Editorial Pòrtic: Barcelona, Spain, 2000.
41. Trapote, M.C.; Rull, V.; Giralt, S.; Corella, J.P.; Montoya, E.; Vegas-Vilarrúbia, T. High-resolution (sub-decadal) pollen analysis of varved sediments from Lake Montcortès (southern Pyrenean flank): A fine-tuned record of landscape dynamics and human impact during the last 500 years. *Rev. Palaeobot. Palynol.* **2018**, *259*, 207–222. [CrossRef]
42. Hakala, A. Meromixis as a part of lake evolution; observations and a revisedclassification of true meromictic lakes in Finland. *Boreal Environ. Res.* **2004**, *9*, 37–53.
43. Jellison, R.; Romero, J.; Melack, J.M. The onset of meromixis during restoration of Mono Lake, California: Unintended consequences of reducingwater diversions. *Limnol. Oceanogr.* **1998**, *43*, 706–711. [CrossRef]
44. Jenny, J.P.; Francus, P.; Normandeau, A.; Lapointe, F.; Perga, M.E.; Ojala, A.; Schimmelmann, A.; Zolitschka, B. Global spread of hypoxia in freshwater ecosystems during the last three centuries is caused by rising local human pressure. *Glob. Chang. Biol.* **2016**, *22*, 1481–1489. [CrossRef]
45. Elliot, J.A. Is the future blue-green? A review of the current model predictions of how climate change could affect pelagic freshwater cyanobacteria. *Water Res.* **2014**, *46*, 1364–1371. [CrossRef] [PubMed]
46. Jankowski, T.; Livingstone, D.M.; Bührer, H.; Forster, R.; Niederhauser, P. Consequences of the 2003 European heat wave for lake temperature profiles, thermal stability, and hypolimnetic oxygen depletion: Implications for a warmer world. *Limnol. Oceanogr.* **2006**, *51*, 815–819. [CrossRef]
47. Rempfer, J.; Livingstone, D.M.; Blodau, C.; Forster, R.; Niederhauser, P.; Kipfer, R. The effect of the exceptionally mild European winter of 2006–2007 on temperature and oxygen profiles in lakes in Switzerland: A foretaste of the future? *Limnol. Oceanogr.* **2010**, *55*, 2170–2180. [CrossRef]

48. Leavitt, P.R. A review of factors that regulate carotenoid and chlorophyll deposition and fossil pigment abundance. *J. Paleolimnol.* **1993**, *9*, 109–127. [CrossRef]
49. Villanueva, J.; Grimalt, J.O.; De Wit, R.; Keely, B.J.; Maxwell, J.R. Sources and transformations of chlorophylls and carotenoids in a monomictic sulphate-rich karstic lake environment. *Org. Geochem.* **1994**, *22*, 739–757. [CrossRef]
50. Hodgson, D.A.; Wright, S.W.; Tyler, P.A.; Davies, N. Analysis of fossil pigments from algae and bacteria in meromictic Lake Fidler, Tasmania, and its application to lake management. *J. Paleolimnol.* **1998**, *19*, 1–22. [CrossRef]
51. North, R.P.; North, R.L.; Livingstone, D.M.; Köster, O.; Kipfer, R. Long-term changes in hypoxia and soluble reactive phosphorus in the hypolimnion of a large temperate lake: Consequences of a climate regime shift. *Glob. Chang. Biol.* **2014**, *20*, 811–823. [CrossRef]
52. Moreno, A.; Giralt, S.; Valero-Garcés, B.; Sáez, A.; Bao, R.; Prego, R.; Pueyo, J.J.; González-Sampériz, P.; Taberner, C. A 14kyr record of the tropical Andes: The Lago Chungará sequence (18° S, northern Chilean Altiplano). *Quat. Int.* **2007**, *161*, 4–21. [CrossRef]
53. Davies, S.J.; Lamb, H.F.; Roberts, S.J. Micro-XRF core scanning in Palaeolimnology:recent developments. In *Micro-XRF Studies of Sediment Cores: Applications of a Non-destructive Tool for the Environmental Sciences*; Croudace, I.W., Rothwell, R.G., Eds.; Springer: Dordrecht, The Netherlands, 2015; pp. 189–226.
54. Camps, J.; Gonzalvo, I.; Güell, J.; López, P.; Tejero, A.; Toldrà, X.; Vallespinós, F.; Vicens, M. El lago de Montcortès, descripción de un ciclo anual. *Oecologia. Aquat.* **1976**, *2*, 99–110.
55. Weijers, J.W.H.; Schouten, S.; van den Donker, J.C.; Hopmans, E.C.; Sinninghe Damsté, J.S. Environmental controls on bacterial tetraether membrane lipid distribution in soils. *Geoch. Cosmoch. Acta* **2007**, *71*, 703–713. [CrossRef]
56. Dang, X.; Yang, H.; Naafs, B.D.A.; Pancost, R.D. Direct evidence of moisture control on the methylation of branched glycerol dialkyl glycerol tetraethers in semi-arid and arid soils. *Geoch. Cosmoch. Acta.* **2016**, *189*, 24–36. [CrossRef]
57. Weijers, J.W.H.; Bernhardt, B.; Peterse, F.; Werne, J.P.; Dungait, J.A.J.; Schouten, S.; Sinninghe Damsté, J.S. Absence of seasonal patterns in MBT-CBT indices in mid-latitude soils. *Geoch. Cosmoch. Acta* **2011**, *75*, 3179–3190. [CrossRef]
58. Bonk, A.; Tylmann, W.; Amann, B.; Enters, D.; Grosjean, M. Modern limnology and varve-formation processes in lake Zabińskie, Northeastern Poland: Comprehensive process studies as a key to understand the sediment record. *J. Limnol.* **2015**, *74*, 358–370. [CrossRef]
59. Miracle, M.R.; Camacho, A.; Juliâ, R.; Vicente, E. Sinking processes and their effect on the sedimentary record in the meromictic Lake La Cruz (Spain). *Verh. Int. Verein. Limnol.* **2000**, *27*, 1209–1213. [CrossRef]
60. Romero, L.; Camacho, A.; Vicente, E.; Miracle, M.R. Sedimentation patterns of photosynthetic bacteria based on pigment markers in meromictic Lake La Cruz (Spain): Paleolimnological implications. *J. Paleolimnol.* **2006**, *35*, 167–177. [CrossRef]
61. Rodrigo, M.A.; Vicente, E.; Miracle, M.R. Short-term calcite precipitation in the karstic meromictic Lake La Cruz (Cuenca, Spain). *Verh. Int. Ver. Limnol.* **1993**, *25*, 711–719. [CrossRef]
62. Rodrigo, M.A.; Vicente, E.; Miracle, M.R. The meromictic Lake La Cruz (Central Spain): Patterns of stratification. *Aquat. Sci.* **2001**, *63*, 406–416. [CrossRef]
63. Maier, D.; Rydberg, J.; Bigler, C.; Renberg, I. Compaction of recent varved lake sediments. *GFF* **2013**, *135*, 231–236. [CrossRef]
64. Butz, C.; Grosjean, M.; Fischer, D.; Wunderle, S.; Tylmann, W.; Rei, B. Hyperspectral imaging spectroscopy: A promising method for the biogeochemical analysis of lake sediments. *J. Appl. Remote Sens.* **2015**, *9*, 1–20. [CrossRef]

quaternary

MDPI

Article

Grain-Size Distribution and Structural Characteristics of Varved Sediments from Lake Żabińskie (Northeastern Poland)

Maurycy Żarczyński [1,*], Jacek Szmańda [2] and Wojciech Tylmann [1]

1 Faculty of Oceanography and Geography, University of Gdansk, 80-309 Gdansk, Poland; wojciech.tylmann@ug.edu.pl
2 Faculty of Geography and Biology, Pedagogical University of Cracow, 30-084 Krakow, Poland; jszmanda@up.krakow.pl
* Correspondence: maurycy.zarczynski@phdstud.ug.edu.pl

Received: 29 November 2018; Accepted: 2 February 2019; Published: 4 February 2019

Abstract: Typically, the description of varve microfacies is based on microscopic sedimentary structures, while standard grain-size analysis is commonly applied with lower resolution. Studies involving a direct comparison of varve microfacies and particle-size distributions, common for clastic environments, are scarce for biogenic varves. In this study, we analyzed nine-year resolution grain-size data from Lake Żabińskie (northeastern Poland) to detect differences between varve microfacies. Six varve microfacies were differentiated using grain-size distributions and sedimentological attributes (calcite layer thickness, dark layer thickness, mass accumulation rate). However, changes in particle-size distributions between different varve types are relatively small and indicate a similar source for the material deposited. Decomposition of grain-size distributions with the end-member approach allows recognition of relative changes for the deposition of allochthonous (mineral) and autochthonous (carbonates, (hydr)oxides) components. Grain-size data suggest that sources of allochthonous material remained constant, while varve formation was controlled mostly by in-lake processes.

Keywords: varves; lake sediments; grain-size; end-members

1. Introduction

Lacustrine sedimentary records are one of the most important paleoenvironmental archives [1]. Among these, records with varved sediments offer a rare opportunity for annually resolved chronologies and analyses [2,3]. Studies based on varved sediments require careful recognition and confirmation of the annual deposition cycle [4,5]. This is mostly achieved with microfacies analysis, which is based on recognition of microscopic sedimentary structures forming the varves [4,6]. Careful studies of present-day in-lake processes might further confirm the annual rhythm of sedimentation [4,7]. Most of the paleoenvironmental analyses rely either on geochemical proxies [8], ecological indicators [9,10] or a combination of those [11]. Textural analysis, including particle-size of the accumulated material, provides one of the fundamental characteristics for the study of sedimentary material [12]. Grain-size analyses offer a variety of research opportunities for lacustrine settings. Sediment texture could be used to study material provenance [13], transport [14], and processes such as paleofloods [15,16], monsoons [17], and aeolian activity [18].

Since the inception of the varve chronology method by De Geer [19], grain-size has played an important, diagnostic role, especially in clastic varves [2]. Textural changes in the cold climate are a result of the freeze-thaw cycles affecting allochthonous material supply [20]. This, in turn, leads to the formation of seasonal laminations, reflecting differences in the composition and size of the accumulated material [3]. Commonly, the warm season is characterized by larger particles, which originate from the catchment, while their transport is induced by snowmelt and increased runoff. During ice-covered seasons, accumulation of coarse particles decreases leading to dominance of smallest particles. This closes the cycle with a fine-grained layer, often with addition of biogenic material [20–22]. Thus, grain-size plays a fundamental role during the recognition of annual cycles. Thanks to continuous density scans and grey-scale image analyses, grain-size data can be obtained with high, seasonal resolution [23,24]. Therefore, this approach is not only valuable for microfacies analysis but even for semi-automatic varve chronology development.

Contrary to clastic or clastic-biogenic varves, mineral matter plays a minor role with biogenic varves that are dominated by material originating from biochemical processes and lacustrine productivity [2]. This causes a lower interest in particle-size research for this type of varves. Different approaches for sediment pretreatment lead to diverse results, due to different components being possibly removed prior to analysis [25]. The most common procedure is to remove organic matter, carbonates and biogenic silica before measurements [14,17,20,26]. However, this leaves possible metal (hydr)oxides intact. This approach could be simplified, for example by removing only organic matter, or further improved with removal of metal (hydr)oxides [25]. Ideally, removal of all non-minerogenic components would effectively lead to the measurement of only allochthonous material of terrestrial origin. However, there is the possibility to destroy clay minerals during subsequent pretreatment steps further influencing the results of measurements. To achieve a full understanding of past processes recorded within biogenic sediments, it might be helpful to measure the size of both allochthonous and autochthonous sediment components.

The aim of this paper is to assess bulk grain-size data as a supporting tool for the discrimination of biogenic varve microfacies. Furthermore, we test the end-member approach for tracing non-detrital sediment components, such as carbonates, metal (hydr)oxides or diatoms. In this study, we used grain-size data from the 2000 year-long, entirely varved sediment core from Lake Żabińskie and compared parameters based on grain-size data with nine-year resolution and previously described varve microfacies.

2. Materials and Methods

2.1. Study Area

Lake Żabińskie (coring site at 54°07′54″ N; 21°59′01.1″ E) is located in the Great Masurian Lakes region (Figure 1A), a part of the Masurian Lakeland (Northeastern Poland) [27]. This postglacial landscape lies within the extent of the Pomeranian Phase of the Vistulian glaciation (ca. 17–16 kBP) [28]. The lake is mostly surrounded by glacial tills, sands, and gravels. Finer material is deposited in depressions and near the lake basins [29,30]. Lake Żabińskie is fed by inflows from Lake Purwin and streams from the south passing arable land. The southernmost tributary originates in the Żabinka village bearing considerable amounts of nutrients and mineral matter, with the latter forming a small delta on the southern shore of the lake [31]. Lake Żabińskie drains to the much larger Lake Gołdopiwo located to the west. The lake basin has a simple morphology with one, centrally located deepest point of 44 m (Figure 1B). Strong seasonal thermal stratification and anoxia in the hypolimnion [4] lead to the formation and preservation of varved lake sediments [31,32].

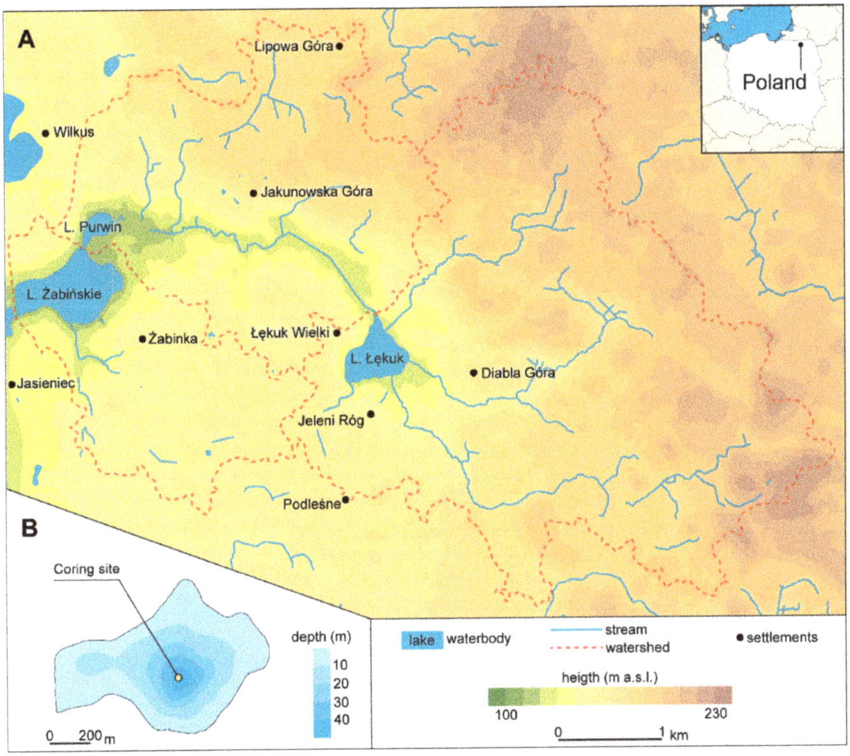

Figure 1. (**A**) Location of Lake Żabińskie. (**B**) Bathymetric map with coring site.

2.2. Coring and Composite Profile

Coring took place during two field campaigns in the summers of 2011 and 2012. A UWITEC (Mondsee, Austria) coring platform was equipped with a 90-mm diameter piston and gravity corers. The complete sediment profile was 19.5 m long, while this study focuses only on the previously investigated top 5.95 m covering the last 2000 years. After retrieval, cores were tightly sealed and stored in a cold room. Later, cores were split into halves, photographed, and macroscopically described. The composite profile was created by correlation of diagnostic horizons.

2.3. Initial Analyses

Core half B was continuously subsampled with a temporal resolution of three years per sample, i.e., an average sample thickness of 8.9 $\pm$ 5.2 mm ($\pm$ 1 SD) with range of 1.6 to 37.0 mm. Dry bulk density was calculated by freeze-drying 1 cm^3 of wet sediment and weighing the dry mass.

2.4. Chronology and Sedimentological Variables

The chronology is based on three individual varve counts and covers 2028 $^{+34}/_{-53}$ varve years, reaching year 17 $^{+34}/_{-53}$ BCE at a depth of 5.95 m. Additional age control of this period was achieved with 29 AMS ^{14}C dates, ^{137}Cs chronostratigraphic horizons, and the Askja cryptotephra of 1875 CE [33]. Details on the chronology are published by Żarczyński et al. [32]. Based on varve chronology, the mass accumulation rate (MAR in g $\times$ cm^{-2} $\times$ $year^{-1}$) was calculated by multiplying varve thickness (VT in mm) by dry bulk density (DBD in g $\times$ cm^{-3}). Additional measurements of calcite layer thickness (CT in mm) and dark layer thickness (DT in mm) were done on scanned images of thin sections (2400 DPI), which were previously used for varve counting. In case of multiple calcite layers within one varve, CT thickness includes the sum of all individual calcite layers occurring in the given year. DT represents all remaining sublayers, and was calculated by subtracting CT from VT.

2.5. Microfacies Analysis

Microfacies analysis was carried out using Axio Imager A2 (ZEISS, Jena, Germany) and L100 (NIKON, Tokyo, Japan) microscopes, with magnifications between 20× and 500×. Thin sections were investigated using non-polarized and polarized light. A detailed overview of the six main microfacies and their formation processes was published with the varve chronology [32].

2.6. Grain-Size Determination

For the determination of particle-size, three parts of equal weight from three consecutive samples were mixed to obtain samples of nine-year resolution. Organic matter was removed by combustion at 550 °C. Carbonates were not removed prior to grain-size measurements. Determination of particle-size distribution was carried out with a Mastersizer 3000 (Malvern, Malvern, United Kingdom) laser particle-size analyzer at the Faculty of Geography and Biology, Pedagogical University of Cracow. Around 50 mg of material were used for each sample to achieve required obscuration of detector. Prior to measurements, ultrasonic disintegration was applied for 60 s. The mean of three measurements for each merged sample was used as the final value.

2.7. Statistics

Raw data is provided as a Supplementary Materials (S1). In a first step, annual sedimentological data for each grain-size sample were averaged into mean values. Because grain-size data are attributed to nine-year samples, this led to more than one varve type per sample. Therefore, the percentage share of a given varve type was calculated for each sample. Only samples with one dominant microfacies (≥70%) were retained for further analyses, excluding 56 (26%) samples from the original dataset of 218 samples. Finally, grain-size data were averaged for each size class to achieve a mean distribution for each varve-microfacies type. Grain-size distribution indices following moment (phi) formulas were calculated with GRADISTAT (version 8.0) [34]. Later, phi values were converted to micrometers. Ternary diagrams were created with ggtern 3.0.0 [35].

Prior to multivariate analyses, variables were log-transformed, scaled and centered with base R (version 3.5.1) [36]. A principal component analysis (PCA) was carried out to determine the relationship between grain-size indices and sedimentological data. Clustering tendencies and differentiation of varve types were visualized on a PCA biplot. To disentangle genetically different material originating from diverse sedimentary processes [37], grain-size distributions were decomposed with an end-member approach using the EMMAgeo (version 0.9.6) package for R [38].

3. Results

3.1. Microfacies

Six microfacies were distinguished for the sediments of Lake Żabińskie (Figure 2; [32]). Type I (68 varves) occurred only at the topmost part of the record and is characterized by well-developed calcite lamina followed by diatoms. The rest of the varve is mostly a mixture of minerogenic and organic detritus with the addition of vivianite and pyrite. Characteristic for this microfacies is the occurrence of additional calcite layers throughout the year. Varve type II (165 varves) characterizes the least distinct annual cycle, with a poorly developed calcite/diatom layer marking the beginning of varve years. Diatom frustules and chrysophyte cysts were deposited on top of the calcite layer, with vivianite and pyrite rarely occurring. A fine-grained mineral layer formed sporadically at the end of the year. Type III (1300 varves) is the most abundant of the record, with the simplest pattern of deposition. The beginning of the varve year is marked by a well-developed calcite layer, followed by deposition of mineral and organic detritus. Chrysophytes were more abundant close to the end of the varve year. Type IV (85 varves) is very similar to type III, with the addition of the thick diatom layer deposited during the second half of the year. Type V (291 varves) starts with a calcite layer enriched with iron, followed by a mix of calcite grains, diatoms, minerogenic and organic matter. Finally, an iron-rich layer was deposited. This diagnostic layer is at times interrupted by a poorly-developed calcite layer. Type VI (118 varves) is similar to type V, with a distinct (thick) diatom layer occurring prior to deposition of the iron-rich layer. Development and occurrence of different microfacies is strongly connected with land-use changes in the catchment and subsequent changes in the lake mixing regime [39]. Varve types III and IV were deposited mostly during an 1100-year long period (ca. 520 to 1620 CE) of catchment reforestation leading to weaker wind action and meromixis (Figure 3). Types I, II, V, and VI occurred during phases of deforestation leading to intensified lake mixing and better oxygenation of the water column. They were influenced by redox processes (types V and VI), enhanced catchment erosion (type II), or recent eutrophication (type I).

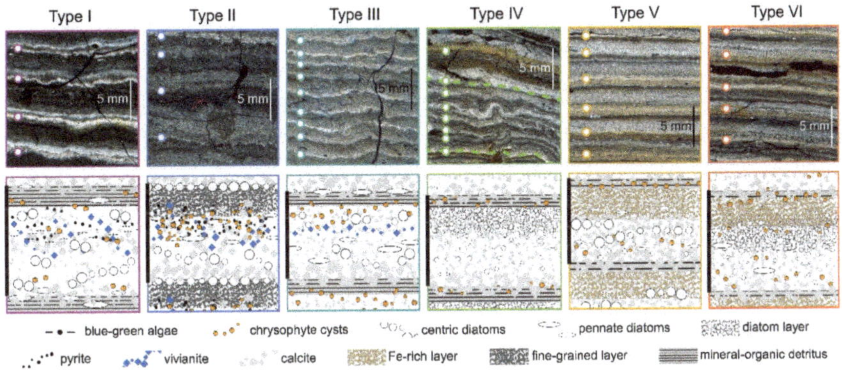

Figure 2. Top: Six varve microfacies (types I–VI) of Lake Żabińskie (thin sections scanned with polarized light). White dots mark the beginning of each varve year. Bottom: Conceptual varve models. Black bars indicate one year.

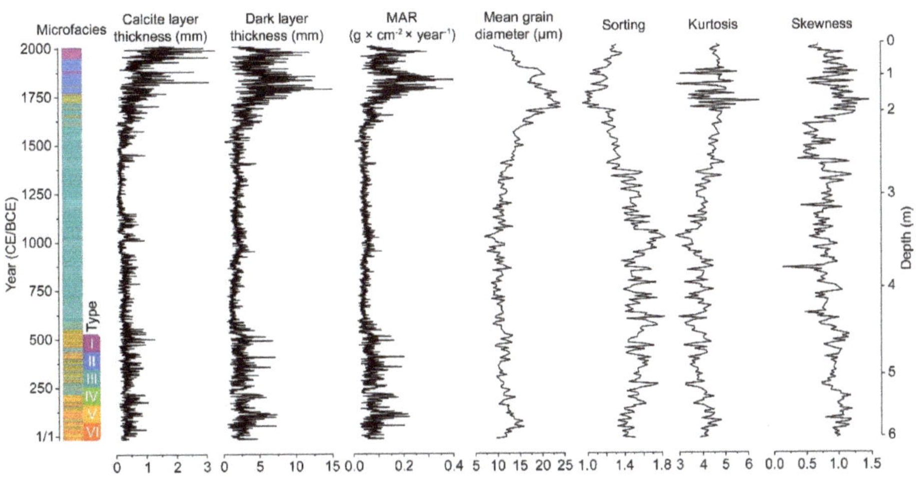

Figure 3. Varve microfacies with sedimentological, and grain-size variability plotted vs. time and depth.

Main patterns of sedimentological properties and grain-size indices are shown in Figure 3 with respective varve microfacies. Overall, both sedimentological characteristics and grain-size parameters are divided into four main phases. The first phase between ca. 17 BCE and 520 CE is characterized by elevated values and higher variability of all the variables except sediment sorting. From 520–1620 CE all values drop and remain relatively stable, with less sediment sorting. Between 1620 and 1870 CE variables rise again, reaching respective maxima with very good sediment sorting. Finally, there is a decoupling between CT and DT, reflected also in mean grain-size with an overall drop toward the top.

After dataset reduction, the number of sediment samples corresponding to each varve type left seven samples of type I, 17 samples of type II, 126 samples of type III, 1 sample of type IV, 8 samples of type V, and 2 samples of type VI. For the needs of statistical analysis, types III and IV as well as V and VI were merged. The main sedimentological properties of the reduced dataset are illustrated by boxplots shown in Figure 4. The mean MAR exceeded 0.10 g $\times$ cm^{-2} $\times$ year^{-1} for most varve types with a maximum of 0.15 $\pm$ 0.05 g $\times$ cm^{-2} $\times$ year^{-1} recorded for type II. The exceptions are types III + IV, with a mean of 0.05 $\pm$ 0.01 g $\times$ cm^{-2} $\times$ year^{-1}. The lowest MAR of 0.03 g $\times$ cm^{-2} $\times$ year^{-1} is recorded within type III + IV, while the maximum value of 0.24 g $\times$ cm^{-2} $\times$ year^{-1} occurred in type II.

Calcite layer thickness varies between varve types, with highest values recorded for type I (1.37 $\pm$ 0.36 mm). The other types were characterized by thinner calcite layers, with type III + IV reaching a minimum of 0.26 $\pm$ 0.10 mm on average. Type III + IV also reaches the lowest value of 0.09 mm, while a maximum of 1.84 mm is recorded within type I. Dark layer thickness shows a similar pattern with highest mean values reached for varve type II (5.43 $\pm$ 1.48 mm), while the lowest value occurs for type III + IV (1.88 $\pm$ 0.51 mm). Overall, varves of types I and II reach higher values for all variables, while type II exhibited the most dispersed properties. Varves of type III + IV are most uniform and generally at the lower end. Type V + VI shows values in between other types.

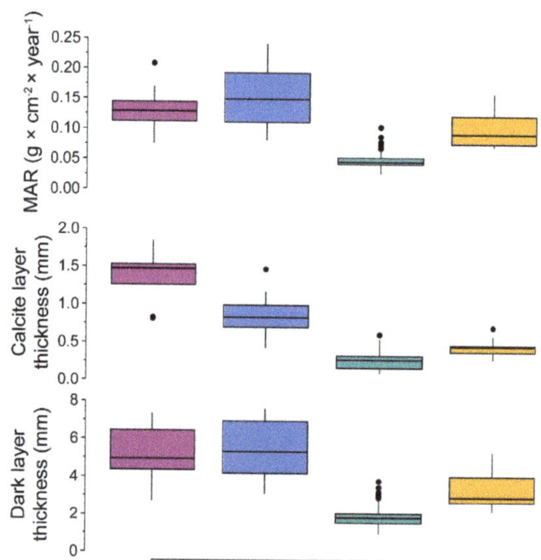

Figure 4. Boxplots of sedimentological properties aggregated for varve types I–VI. Boxplot hinges represent the 25th and 75th quantiles, solid horizontal line represents the mean, notches indicate maximum and minimum values within the 1.5 inter-quartile range and black dots represent outliers.

3.2. *Grain-Size*

Type II varves are dominated by relatively coarse material with a mean particle-size of 13.5–20.9 µm. Lower mean particle-size is associated with type III + IV (7.1–24.3 µm). Overall, types I, III + IV and V + VI are characterized by a minimum mean particle-size below 10 µm. On the other hand, the maximum mean particle-size exceeds 20.0 μm within types II, III + IV, and V + VI. The median particle-size, together with the 10th (P_{10}) and 90th (P_{90}) percentiles show the margins of distributions and follow similar patterns like the mean particle-size (Figure 5a). The mean kurtosis ranges from 4.01 (type III + IV) to 4.66 (type II). Mean skewness is between 0.78 (type III + IV) and 1.07 (type V + VI), with type V + VI reaching overall higher values and low dispersions.

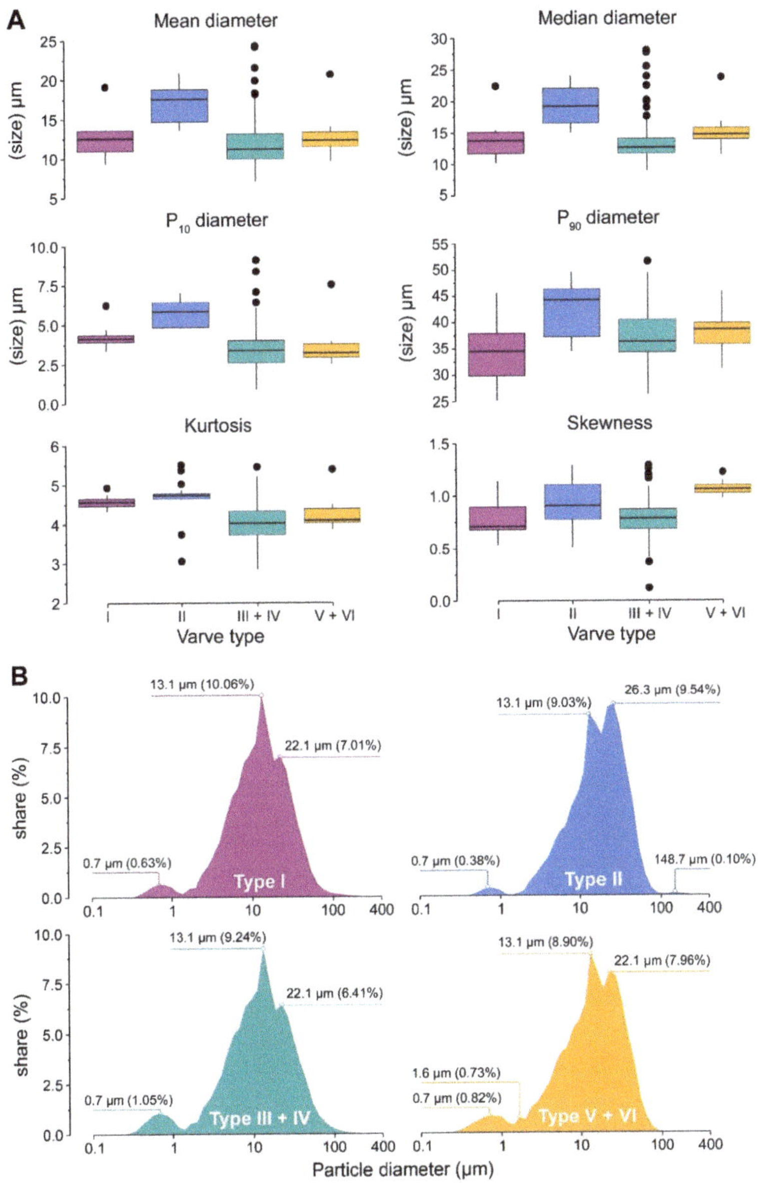

Figure 5. (**a**) Boxplots of main grain-size indices and parameters aggregated by varve type. (**b**) Average grain-size distribution aggregated by varve type (labels indicate positions of modes, respective size, and relative share). Boxplots features as in Figure 4.

Averaged distributions of particle-size, aggregated by type (Figure 5b), show that varve types II and V + VI are in fact shifted towards coarser material relative to the other types. Type I shows a similar distribution like type III + IV. However, the secondary grain-size peak is smaller than 1 μm and better pronounced for type III + IV than for type I. While all the distributions are shifted towards coarser grains, the secondary peak below 1 μm is also well-pronounced. Grain-size distributions of all types are determined by a dominant silt fraction. Main shifts occur for coarser silt around 22.1 μm (types I, III + IV, V + VI) or 26.3 μm (type II), and finer silt around 13.1 μm (all types). Mean grain-sizes calculated for average distributions range from 11.5 μm (type III + IV) to 16.9 μm (type II), with three to four modes (Figure 5b) appearing mostly at 22.1 μm, 13.1 μm, and 0.7 μm. Characteristic for type II is an additional, coarse mode at 148.7 μm. The samples were mostly poorly (n = 159) or moderately (n = 3) sorted, with either meso (n = 31) or leptokurtic (n = 131), fine (n = 159) skewed or symmetrical (n = 3) distribution.

Overall, all varves belong either to medium silt (I and III + IV) or coarse silt (II and V + VI) grain-size classes. The relationships between different sediment classes are shown in Figure 6. Some degree of separation is visible between types, with type III + IV shifted slightly to the direction of clay and sand-sized particles, with a gradual change towards silt-dominated type II. The mean clay content varies between 1.7 ± 0.8% (type II) and 6.3 ± 3.3% (type III + IV). Silt dominates all samples with rather high differences between types. It is least abundant in type III + IV with a mean of 91.2 ± 3.3%, while for type II silt reaches 96.4 ± 1.1%. Sand-sized particles are most prominent in type III + IV (2.4 ± 1.2%).

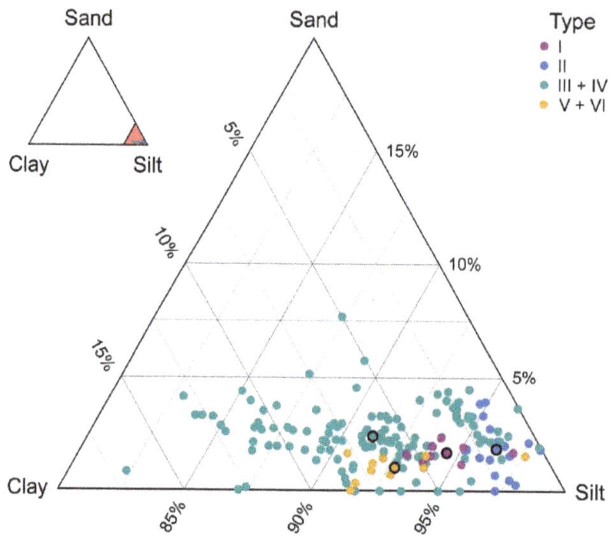

Figure 6. Ternary diagram of grain-size data. Larger points indicate values for mean grain-size distributions. The smaller triangle shows the location of samples within the full ternary diagram.

3.3. *Statistical Analyses*

3.3.1. Principal Components Analysis

Principal components analysis (PCA) was carried out on three main sedimentological variables (MAR, CT, DT) with the addition of parameters calculated from grain-size distributions (Figure 7). Four principal components exceeded 5% of the total explained variance (PC1: 57.2%; PC2: 16.4%; PC3: 11.3%; PC4: 8.6%), while only PC1 passed the broken-stick test. There are three main groups of visible vectors: (1) sedimentological variables (CT, DT, MAR) and skewness, (2) variables describing distribution (P_{10}, P_{90}, mean, and median diameter) with kurtosis, and (3) sorting. Samples tended to cluster to some extent: types I and II were mostly related to sedimentological variables while type III + IV had strong relationships with variables describing particle-size distributions (except skewness). Samples from type V + VI were more dispersed between other types.

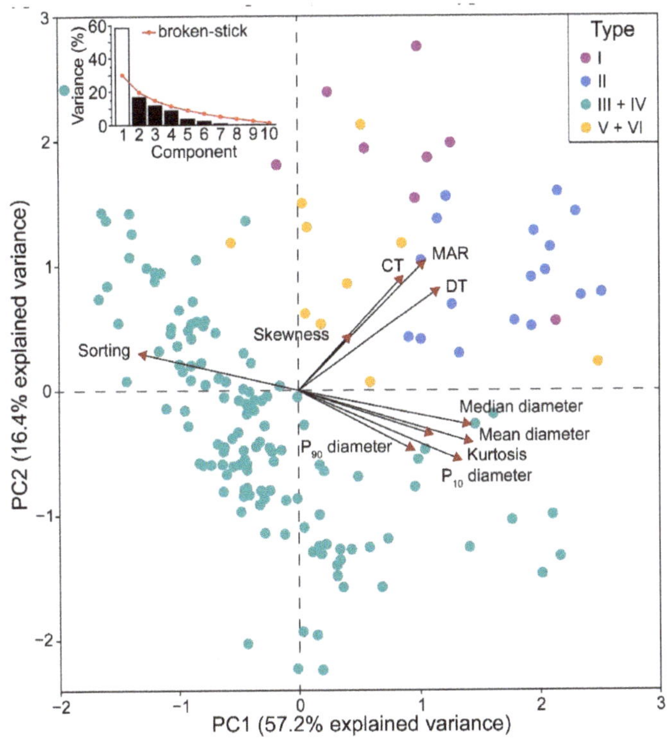

Figure 7. PCA biplot of sedimentological and grain-size data. CT—calcite layer thickness; DT—dark layer thickness; MAR—mass accumulation rate.

3.3.2. End-Member Analysis

End-member analysis resulted in recognition of five different end-members (EM) (Figure 8a). Coarsest end-members explain the smallest proportion of variation with only 5.4% and 5.0% for EM1 (mode at 44.0 μm) and EM2 (mode at 37.0 μm), respectively. The end-members associated with finer material (7.8 μm to 31.0 μm) explain over 25% of variance each. EM1 was mostly visible within type III + IV, with some observations occurring within type II and IV. EM2 was recorded only in type III + IV, rarely dominating this type. EM3 was highly variable among the types, with highest relative values recorded within type II and V + VI. Types I and III + IV reach overall low values, with some outliers in type III + IV. EM4 shows a gradual decrease starting from type I until type III + IV. A slight rise within type V + VI is also visible. EM5 is virtually absent within type II and reaches the highest values within type III + IV.

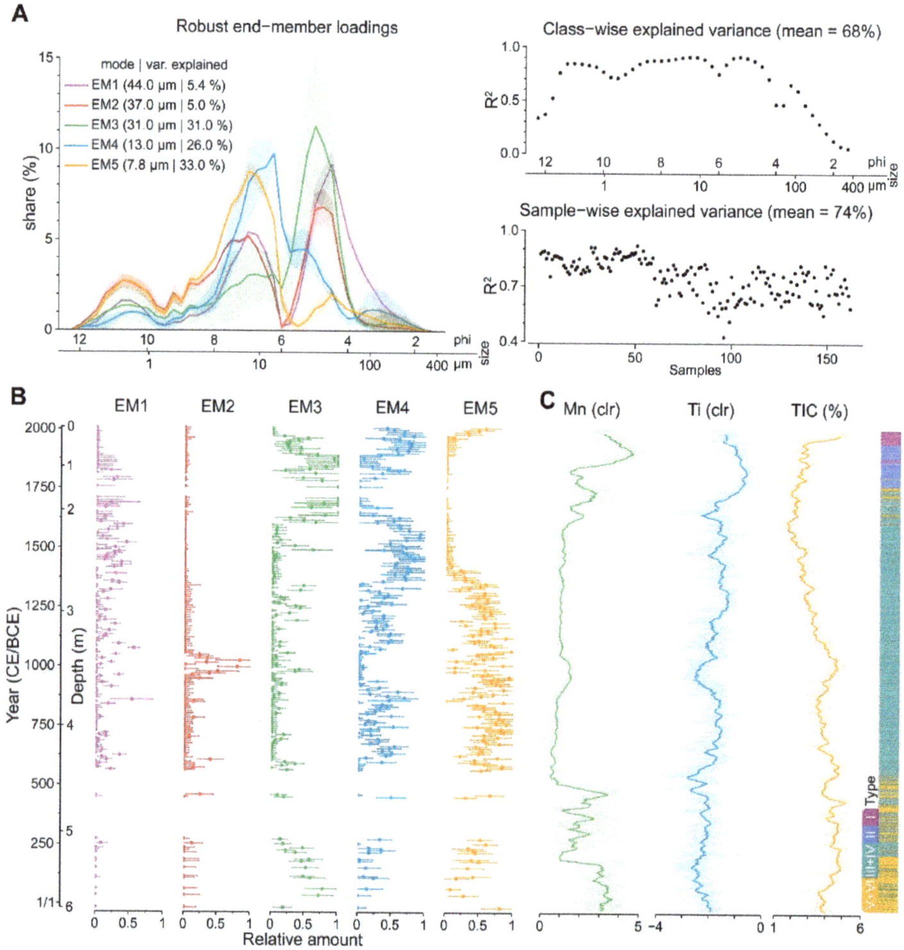

Figure 8. (**a**) End-members loadings with diagnostic plots. (**b**) Relative abundance of end-members vs time and depth. (**c**) Profiles of selected geochemical proxies (µXRF centered log-ratios (clr) of Mn and Ti, TIC—total inorganic carbon) with a 20-year running mean and varve types (Żarczyński et al., 2019; modified).

4. Discussion

Biogenic varves from Lake Żabińskie show macro- and microscopic differences, mostly resulting from variations of in-lake processes. Despite different structures, the texture of these sediments is generally similar and, as a result of their autochthonous origin, dominated by silt-sized particles. This can be related to calcite crystals as well as diatom frustules abundant in all sediment samples. Sediments of Lake Żabińskie are dominated by small (approx. 4 to 50 µm), planktonic diatom taxa (*Aulacoseira, Stephanodiscus, Cyclotella*). Larger (over 100 µm), benthic diatoms (*Fragilaria, Pseudostaurosira, Staurosira*) are scarce. Therefore, inferences about external transport processes and sources are limited due to minor contributions of allochthonous material associated with aeolian or fluvial provenance. Annually resolved sedimentological data, combined with the standard grain-size allows separation of Lake Żabińskie varve types. Differences based on mean, median, P_{10}, and P_{90} particle-sizes show similar patterns. Type II varves consistently document the highest values, suggesting contributions linked to relatively coarser material showing elevated erosion rates [40]. On the other hand, the remaining varve types exhibit similar patterns, mostly due to deposition of autochthonous material. However, type III + IV varves have numerous outliers of coarser particle-size. This is probably due to episodic deposition of aggregated autochthonous material such as calcite [41].

In terms of statistical measures such as sorting, skewness and kurtosis, all varve types exhibit considerable similarity. Analyzed sediments are poorly sorted due to dominance of the silt fraction, which is deposited under lower flow conditions [42,43], which do not favor sorting processes. Transport and deposition of sediments in the lakes could be directly influenced by wind fetch and wind stress, effectively changing sedimentation patterns over time, potentially leading to complex sorting processes [44]. Furthermore, sediments of Lake Żabińskie are comprised of products of biogenic (diatoms) and biochemical (carbonates, Fe and Mn (hydr)oxides) processes, which are more important than transport and sorting. Selective sediment deposition is demonstrated by positive skewness associated with dominance of a fraction coarser than the median [45]. On the other hand, highest skewness values recorded within type II show a more diverse deposition setting due to the higher influence of allochthonous material. Grain-size distributions of varve types do not differ significantly, and fraction modes are similar, with minor changes of their relative shares (Figure 5b).

Additionally, PCA analysis of grain-size and sedimentological data from Lake Żabińskie results to some degree in separation of different varve types. The close relationship between CT, MAR and type I varves suggests thickness of the calcite layer as the controlling factor for the overall accumulation of the given year—a possible response to the increasing lacustrine productivity [46] and carbonate deposition [47]. On the other hand, type II varves are associated more with DT implying higher importance of accumulation of detrital components rather than poorly developed calcite layers. These varves developed in response to forest clearing in the catchment area. Mineral matter deposition was caused by soil erosion and increased delivery of allochthonous material [40], leading to a more prominent role of terrestrial matter in the varve structure. Additionally, at that time intensity of lake mixing increased, which caused poor varve preservation. Varve type III + IV is related to major grain-size characteristics, i.e., sorting or mean particle-size and kurtosis. This type is characteristic for an over 1000 year-long period of meromixis, when catchment was densely forested reducing wind stress on the lake [39]. Type V + VI shows no apparent tendencies, which are explained by the complicated structure and origin, where both sedimentation and chemical processes played important roles. Occurrence of this type is characteristic for periods of more intense lake mixing. This in turn caused deposition of Fe and Mn (hydr)oxides forming visible sublayers in varves. These varves co-occurred with early human impact on the lake catchment [39].

End-member analysis and distribution decomposition allows separation and recognition of the sources and transport processes of allochthonous material. As allochthonous input is of minor importance for Lake Żabińskie, we attempt to compare end-members with geochemical data (Figure 8c) [11,39]. This leads to the recognition and definition of non-clastic end-members. Significance of correlations is corrected for autocorrelation [48]. The pattern visible in EM3 resembles Mn variability of µXRF data ($r = 0.46$, $p < 0.05$). With a mode around 31.0 µm, this indicates presence of cemented aggregates of mineral matter or carbonates coated with these (hydr)oxides. Mn and Fe variability in sediments of Lake Żabińskie during the last 2000 years was controlled mostly by changes in oxygenation of the water column [39]. Higher abundance in the type V + VI seems to support this interpretation. The high values noted also within type II varves are a sign of metal scavenging, adsorption onto silt-sized mineral particles and coating of grains [25]. Comparing EM4 with the Ti record ($r = 0.35$, $p < 0.05$) suggests deposition of silt-sized (mode at 13.0 µm) allochthonous mineral matter. EM5 is probably associated with the abundance of calcite in the sediments showing a similar pattern like TIC [39] ($r = 0.67$, $p < 0.05$). Calcite is an important component for every type, except type II, where calcite laminas are poorly developed and of low quality [40]. EM1 and EM2 explain a relatively small portion of the total variance and complicate interpretations. Finally, superimposed peaks of end-member distributions might be caused by the authigenic nature of the sediments. Precipitation and redeposition processes could lead to a different size of accumulated material of the same origin. Furthermore, the end-member analysis should be treated with caution, especially when run without reference based on samples of known origin [49]. Analysis of different algorithms conducted by van Hateren et al. [50] suggests that EMMAgeo might be prone to the occurrence of artifacts in end-member grain-size distributions. Yet, comparison with geochemical data might act as external validation.

It should be stressed that there are multiple sources of uncertainty in our analysis. The first is related to the inhomogeneity of samples that contain more than one varve type. In this way, signals from two varve types are mixed, leading to a weak separation between grain-size characteristics of different varve types. Secondly, lacustrine sedimentation in low-land lakes of the temperate zone is dominated by in-lake production and biogeochemical processes rather than accumulation of minerogenic matter. Due to this, typical discrimination between transport regimes and particle deposition could not be applied. The finest material tends to aggregate when settling down the water column, which is important for metal compounds and (hydr)oxides or carbonates [51]. Additionally, even though not studied here, diatoms further influence grain-size determination. Diatom frustules often are abundant and come in a variety of shapes and sizes. These factors influence the interpretation when the range of particle-size is narrow and covers more than purely allochthonous matter.

Our results suggest that even bulk data of lower, non-annual resolution is useful for discrimination between varve types. Comparison of structure and aggregated grain-size analyses improves and validates previously determined varve microfacies. A step forward will be varve-by-varve sampling, sequential extraction of sediment components and repeated grain-size measurements. If applied, this might lead to recognition of particle-sizes associated with different components. Furthermore, it would be helpful in assessing the relation between particle-size and their scavenging potential [52,53].

5. Conclusions

We compared bulk grain-sized data with an annually resolved record of varve properties and microfacies and found that even with aggregated data it is possible to partially discriminate varve microfacies. End-member analysis resulted in the attribution of end-members to specific sedimentation styles (i.e., precipitation of carbonates, formation of metal (hydr)oxides, and deposition of clastic particles), as well as tracing relative changes of their respective abundance over time. Therefore, we conclude that grain-size analysis is useful for more than characterization of the mineral component of biogenic lake sediments. Further studies with sequential extraction and separation of individual sediment components will likely improve the interpretation of end-member analysis.

Supplementary Materials: The following are available online at http://www.mdpi.com/2571-550X/2/1/8/s1.

Author Contributions: Conceptualization: M.Ż. and W.T.; methodology: M.Ż., W.T., and J.S.; investigation: M.Ż.; writing—original draft preparation: M.Ż. and W.T.; writing—review and editing: M.Ż., W.T., and J.S.; visualization: M.Ż.; project administration: W.T.; funding acquisition: W.T. and M.Ż.

Funding: This research was funded by National Science Centre, grant number 2014/13/B/ST10/01311. Additionally, MZ was funded by University of Gdansk, grant number 538-G110-B576-16.

Acknowledgments: We would like to sincerely thank Anna Poraj-Górska, Dorota Chmielowska, Michał Marosz, and Benjamin Amann for their help and contribution. We also thank three anonymous reviewers for their comments which substantially improved this manuscript.

Conflicts of Interest: The authors declare no conflict of interest.

References

1. Brauer, A. Annually Laminated Lake Sediments and Their Palaeoclimatic Relevance. In *The Climate in Historical Times: Towards a Synthesis of Holocene Proxy Data and Climate Models*; Fischer, H., Kumke, T., Lohmann, G., Flöser, G., Miller, H., von Storch, H., Negendank, J.F.W., Eds.; Springer: Berlin/Heidelberg, Germany, 2004; Volume 18, pp. 109–127, ISBN 978-3-662-10313-5.

2. Zolitschka, B.; Francus, P.; Ojala, A.E.K.; Schimmelmann, A. Varves in lake sediments—A review. *Quat. Sci. Rev.* **2015**, *117*, 1–41. [CrossRef]
3. Ojala, A.E.K.; Francus, P.; Zolitschka, B.; Besonen, M.; Lamoureux, S.F. Characteristics of sedimentary varve chronologies—A review. *Quat. Sci. Rev.* **2012**, *43*, 45–60. [CrossRef]
4. Bonk, A.; Tylmann, W.; Amann, B.; Enters, D.; Grosjean, M. Modern limnology, sediment accumulation and varve formation processes in Lake Żabińskie, northeastern Poland: Comprehensive process studies as a key to understand the sediment record. *J. Limnol.* **2015**, *74*, 358–370. [CrossRef]
5. Zolitschka, B. Varved lake sediments. In *Encyclopedia of Quaternary Science*; Elsevier B.V.: Amsterdam, The Netherlands, 2007; pp. 3105–3114, ISBN 978-0-444-52747-9.
6. Dräger, N.; Theuerkauf, M.; Szeroczyńska, K.; Wulf, S.; Tjallingii, R.; Plessen, B.; Kienel, U.; Brauer, A. Varve microfacies and varve preservation record of climate change and human impact for the last 6000 years at Lake Tiefer See (NE Germany). *Holocene* **2017**, *27*, 450–464. [CrossRef]
7. Trapote, M.C.; Vegas-Vilarrúbia, T.; López, P.; Puche, E.; Gomà, J.; Buchaca, T.; Cañellas-Boltà, N.; Safont, E.; Corella, J.P.; Rull, V. Modern sedimentary analogues and integrated monitoring to understand varve formation in the Mediterranean Lake Montcortès (Central Pyrenees, Spain). *Palaeogeogr. Palaeoclimatol. Palaeoecol.* **2018**, *496*, 292–304. [CrossRef]
8. Last, W.M.; Smol, J.P. An Introduction to Physical and Geochemicalmethods Used in Paleolimnology. In *Tracking Environmental Change Using Lake Sediments. Volume 2: Physical and Geochemical Methods*; Last, W.M., Smol, J.P., Eds.; Kluwer Academic Publishers: Dordrecht, The Netherlands, 2001; Volume 2, pp. 1–5, ISBN 978-0-306-47670-9.
9. Smol, J.P.; Birks, J.; Last, W.M. Using biology to study long-term environmental change. In *Tracking Environmental Change Using Lake Sediments. Volume 3: Terrestrial, Algal, and Siliceous Indicators*; Smol, J.P., Birks, H.J.B., Last, W.M., Eds.; Kluwer Academic Publishers: Dordrecht, The Netherlands, 2001; Volume 3, p. 3.
10. Smol, J.P.; Birks, H.J.B.; Last, W.M. Zoological indicators in lake sediments: An introduction. In *Tracking Environmental Change Using Lake Sediments. Volume 4: Zoological Indicators*; Smol, J.P., Birks, H.J.B., Last, W.M., Eds.; Kluwer Academic Publishers: Dordrecht, The Netherlands, 2001; Volume 4, pp. 1–4.
11. Hernández-Almeida, I.; Grosjean, M.; Gomez-Navarro, J.J.; Larocque-Tobler, I.; Bonk, A.; Enters, D.; Ustrzycka, A.; Piotrowska, N.; Przybylak, R.; Wacnik, A.; et al. Resilience, rapid transitions and regime shifts: Fingerprinting the responses of Lake Zabinskie (NE Poland) to climate variability and human disturbance since AD 1000. *Holocene* **2017**, *27*, 258–270. [CrossRef]
12. Last, W.M. Textural Analysis of Lake Sediments. In *Tracking Environmental Change Using Lake Sediments. Volume 2: Physical and Geochemical Methods*; Last, W.M., Smol, J.P., Eds.; Kluwer Academic Publishers: Dordrecht, The Netherlands, 2001; Volume 2, pp. 41–81, ISBN 978-0-306-47670-9.
13. Jin, Z.; Li, F.; Cao, J.; Wang, S.; Yu, J. Geochemistry of Daihai Lake sediments, Inner Mongolia, north China: Implications for provenance, sedimentary sorting, and catchment weathering. *Geomorphology* **2006**, *80*, 147–163. [CrossRef]
14. IJmker, J.; Stauch, G.; Dietze, E.; Hartmann, K.; Diekmann, B.; Lockot, G.; Opitz, S.; Wünnemann, B.; Lehmkuhl, F. Characterisation of transport processes and sedimentary deposits by statistical end-member mixing analysis of terrestrial sediments in the Donggi Cona lake catchment, NE Tibetan Plateau. *Sediment. Geol.* **2012**, *281*, 166–179. [CrossRef]
15. Schillereff, D.N.; Chiverrell, R.C.; Macdonald, N.; Hooke, J.M. Flood stratigraphies in lake sediments: A review. *Earth-Sci. Rev.* **2014**, *135*, 17–37. [CrossRef]
16. Ota, Y.; Kawahata, H.; Sato, T.; Seto, K. Flooding history of Lake Nakaumi, western Japan, inferred from sediment records spanning the past 700 years. *J. Quat. Sci.* **2017**, *32*, 1063–1074. [CrossRef]
17. Liu, X.; Vandenberghe, J.; An, Z.; Li, Y.; Jin, Z.; Dong, J.; Sun, Y. Grain size of Lake Qinghai sediments: Implications for riverine input and Holocene monsoon variability. *Palaeogeogr. Palaeoclimatol. Palaeoecol.* **2016**, *449*, 41–51. [CrossRef]
18. Zhai, Q.; Guo, Z.; Li, Y.; Li, R. Annually laminated lake sediments and environmental changes in Bashang Plateau, North China. *Palaeogeogr. Palaeoclimatol. Palaeoecol.* **2006**, *241*, 95–102. [CrossRef]
19. De Geer, G. On late Quaternary time and climate. *Geol. Foereningan i Stock. Foerhandlingar* **1908**, *30*, 459–464. [CrossRef]

20. Ojala, A.E.K.; Kosonen, E.; Weckström, J.; Korkonen, S.; Korhola, A. Seasonal formation of clastic-biogenic varves: The potential for palaeoenvironmental interpretations. *GFF* **2013**, *135*, 237–247. [CrossRef]
21. Haltia-Hovi, E.; Saarinen, T.; Kukkonen, M. A 2000-year record of solar forcing on varved lake sediment in eastern Finland. *Quat. Sci. Rev.* **2007**, *26*, 678–689. [CrossRef]
22. Amann, B.; Lamoureux, S.F.; Boreux, M.P. Winter temperature conditions (1670–2010) reconstructed from varved sediments, western Canadian High Arctic. *Quat. Sci. Rev.* **2017**, *172*, 1–14. [CrossRef]
23. Ojala, A.E.K.; Francus, P. Comparing X-ray densitometry and BSE-image analysis of thin section in varved sediments. *Boreas* **2002**, *31*, 57–64. [CrossRef]
24. Ojala, A.E.K. Application of X-ray radiography and densitometry in varve analysis. In *Image Analysis, Sediments and Paleoenvironments*; Francus, P., Ed.; Springer: Dordrecht, The Netherlands, 2004; pp. 187–202.
25. Gee, G.W.; Or, D. Particle-Size Analysis. In *Methods of Soil Analysis: Part 4 Physical Methods*; SSSA Book Ser. 5.4; Soil Science Society of America: Madison, WI, USA, 2002; pp. 255–293, ISBN 978-0-89118-893-3.
26. Augustinus, P.; Bleakley, N.; Deng, Y.; Shane, P.; Cochran, U. Rapid change in early Holocene environments inferred from Lake Pupuke, Auckland City, New Zealand. *J. Quat. Sci.* **2008**, *23*, 435–447. [CrossRef]
27. Solon, J.; Borzyszkowski, J.; Bidłasik, M.; Richling, A.; Badora, K.; Balon, J.; Brzezińska-Wójcik, T.; Chabudziński, Ł.; Dobrowolski, R.; Grzegorczyk, I.; et al. Physico-geographical mesoregions of Poland: Verification and adjustment of boundaries on the basis of contemporary spatial data. *Geogr. Pol.* **2018**, *91*, 143–170. [CrossRef]
28. Marks, L.; Dzierżek, J.; Janiszewski, R.; Kaczorowski, J.; Lindner, L.; Majecka, A.; Makos, M.; Szymanek, M.; Tołoczko-Pasek, A.; Woronko, B. Quaternary stratigraphy and palaeogeography of Poland. *Acta Geol. Pol.* **2016**, *66*, 403–427. [CrossRef]
29. Szumański, A. *Objaśnienia do Szczegółowej Mapy Geologicznej Polski, Arkusz Giżycko (104) (Explanation to the Detailed Geological Map of Poland, Sheet Giżycko (104))*; Państwowy Instytut Geologiczny: Warszawa, Poland, 2000.
30. Pochocka-Szwarc, K.; Lisicki, S. *Objaśnienia do Szczegółowej Mapy Geologicznej Polski, Arkusz Orłowo (105) (Explanation to the Detailed Geological Map of Poland, Sheet Orłowo (105))*; Państwowy Instytut Geologiczny—Państwowy Instytut Badawczy: Warszawa, Poland, 2015; ISBN 9788378634331.
31. Bonk, A.; Tylmann, W.; Goslar, T.; Wacnik, A.; Grosjean, M. Comparing Varve Counting And 14C-Ams Chronologies In The Sediments Of Lake Żabińskie, Northeastern Poland: Implications For Accurate 14C Dating Of Lake Sediments. *Geochronometria* **2015**, *42*, 159–171. [CrossRef]
32. Żarczyński, M.; Tylmann, W.; Goslar, T. Multiple varve chronologies for the last 2000 years from the sediments of Lake Żabińskie (northeastern Poland)—Comparison of strategies for varve counting and uncertainty estimations. *Quat. Geochronol.* **2018**, *47*, 107–119. [CrossRef]
33. Tylmann, W.; Bonk, A.; Goslar, T.; Wulf, S.; Grosjean, M. Calibrating 210Pb dating results with varve chronology and independent chronostratigraphic markers: Problems and implications. *Quat. Geochronol.* **2016**, *32*, 1–10. [CrossRef]
34. Blott, S.J.; Pye, K. GRADISTAT: A grain size distribution and statistics package for the analysis of unconsolidated sediments. *Earth Surf. Process. Landf.* **2001**, *1248*, 1237–1248. [CrossRef]
35. Hamilton, N. ggtern: An Extension to "ggplot2", for the Creation of Ternary Diagrams 2018. Available online: https://cran.r-project.org/package=ggtern (accessed on 1 November 2018).
36. R Core Team. *R: A Language and Environment for Statistical Computing*; R Core Team: Vienna, Austria, 2018.
37. Dietze, E.; Maussion, F.; Ahlborn, M.; Diekmann, B.; Hartmann, K.; Henkel, K.; Kasper, T.; Lockot, G.; Opitz, S.; Haberzettl, T. Sediment transport processes across the Tibetan Plateau inferred from robust grain-size end members in lake sediments. *Clim. Past* **2014**, *10*, 91–106. [CrossRef]
38. Dietze, M.; Dietze, E. EMMAgeo: End-Member Modelling of Grain-Size Data 2016. Available online: https://github.com/coffeemuggler/EMMAgeo (accessed on 1 November 2018).
39. Żarczyński, M.; Wacnik, A.; Tylmann, W. Tracing lake mixing and oxygenation regime using the Fe/Mn ratio in varved sediments: 2000 year-long record of human-induced changes from Lake Żabińskie (NE Poland). *Sci. Total Environ.* **2019**, *657*, 585–596. [CrossRef] [PubMed]
40. Bonk, A.; Kinder, M.; Enters, D.; Grosjean, M.; Meyer-Jacob, C.; Tylmann, W. Sedimentological and geochemical responses of Lake Żabińskie (north-eastern Poland) to erosion changes during the last millennium. *J. Paleolimnol.* **2016**, *56*, 239–252. [CrossRef]

41. Czymzik, M.; Dulski, P.; Plessen, B.; Von Grafenstein, U.; Naumann, R.; Brauer, A. A 450 year record of spring-summer flood layers in annually laminated sediments from Lake Ammersee (southern Germany). *Water Resour. Res.* **2010**, *46*, 1–16. [CrossRef]
42. Sly, P.G.; Thomas, R.L.; Pelletier, B.R. Interpretation of moment measures derived from water-lain sediments. *Sedimentology* **1983**, *30*, 219–233. [CrossRef]
43. Szmańda, J.B. Main determinants of the grain size distribution of overbank deposits in Poland—An overview of literature on models of sedimentation. *Geol. Q.* **2018**, *62*, 873–880. [CrossRef]
44. Håkanson, L. The influence of wind, fetch, and water depth on the distribution of sediments in Lake Vänern, Sweden. *Can. J. Earth Sci.* **1977**, *14*, 397–412. [CrossRef]
45. McLaren, P. An Interpretation of Trends in Grain Size Measures. *J. Sediment. Res.* **1981**, *51*, 629–714. [CrossRef]
46. Amann, B.; Lobsiger, S.; Tylmann, W.; Filipiak, J.; Grosjean, M. Spring temperature variability and eutrophication history inferred from sedimentary pigments in the varved sediments of Lake Zabinskie, northeast Poland, 1907-2008 AD. *Glob. Planet. Chang.* **2014**, *123*, 86–96. [CrossRef]
47. Dean, W.E.; Megard, R.O. Environment of deposition of CaCO3 in Elk Lake, Minnesota. In *Elk Lake, Minnesota: Evidence for Rapid Climate Change in the North-Central United States*; Bradbury, J.P., Dean, W.E., Eds.; Geological Society of America: Boulder, CO, USA, 1993; pp. 97–114.
48. Bayley, G.V.; Hammersley, J.M. The "Effective" Number of Independent Observations in an Autocorrelated Time Series. *Suppl. J. R. Stat. Soc.* **1946**, *8*, 184. [CrossRef]
49. Weltje, G.J.; Prins, M.A. Muddled or mixed? Inferring palaeoclimate from size distributions of deep-sea clastics. *Sediment. Geol.* **2003**, *162*, 39–62. [CrossRef]
50. van Hateren, J.A.; Prins, M.A.; van Balen, R.T. On the genetically meaningful decomposition of grain-size distributions: A comparison of different end-member modelling algorithms. *Sediment. Geol.* **2018**, *375*, 49–71. [CrossRef]
51. Leng, M.J.; Jones, M.D.; Frogley, M.R.; Eastwood, W.J.; Kendrick, C.P.; Roberts, C.N. Detrital carbonate influences on bulk oxygen and carbon isotope composition of lacustrine sediments from the Mediterranean. *Glob. Planet. Chang.* **2010**, *71*, 175–182. [CrossRef]
52. Filipek, L.H.; Owen, R.M. Geochemical associations and grain-size partitioning of heavy metals in lacustrine sediments. *Chem. Geol.* **1979**, *26*, 105–117. [CrossRef]
53. Horowitz, A.J.; Elrick, K.A. The relation of stream sediment surface area, grain size and composition to trace element chemistry. *Appl. Geochem.* **1987**, *2*, 437–451. [CrossRef]

MDPI
St. Alban-Anlage 66
4052 Basel
Switzerland
Tel. +41 61 683 77 34
Fax +41 61 302 89 18
www.mdpi.com

Quaternary Editorial Office
E-mail: quaternary@mdpi.com
www.mdpi.com/journal/quaternary